ALLE ZEIT WACH
1842

Louis A. Pilato · Michael J. Michno

Advanced Composite Materials

With 50 Figures

Springer-Verlag
Berlin Heidelberg New York
London Paris Tokyo
Hong Kong Barcelona Budapest

Dr. Louis A. Pilato

598 Watchung Road
Bound Brook, NJ 08805
USA

Michael J. Michno

Amoco Performance Products Inc.
Atlanta, GA,
USA

ISBN 3-540-57563-4 Springer-Verlag Berlin Heidelberg New York
ISBN 0-387-57563-4 Springer-Verlag New York Berlin Heidelberg

Library of Congress Cataloging-in-Publication Data
Pilato, L. (Louis), 1934 – Advanced composite materials / Louis A. Pilato, Michael J. Michno. p.cm.
Includes index.
ISBN 3-540-57563-4.
ISBN 0-387-57563-4 (U.S.)
1. Polymeric composites. I. Michno, Michael J., 1944 – II. Title.

Production: PRODUserv Springer Produktions-Gesellschaft, Berlin;
Cover: Erich Kirchner, Heidelberg; Typesetting: Macmillan India Ltd., Bangalore
SPIN: 10025874 53/3020 – 5 4 3 2 1 0 – Printed on acid-free paper

Preface

Advanced Composite Materials or polymeric matrix based composites is the primary topic treated in this book. Thirty years have elapsed since the performance characteristics of these composites fostered their use in aircraft component parts. Today advanced composite materials are embraced by the end-user community as viable high performance engineering materials for demanding applications.

Acceptance of composites and growth in applications required acceptance and understanding by many functional groups, e.g., designers, materials and process engineers, quality assurance specialists, purchasing agents and ultimately customers. The role of the customer for aircraft component parts is just as important but not as easily recognizable as the role of the customer in the sporting goods industry. Advanced composite materials have revolutionized sports equipment where, in fact, these novel high performance materials have displaced the traditional materials of rudimentary sports models.

Similarly the evolution of new composite systems has required cross functional or multi-disciplinary understanding and contributions by those engaged in this technology intensive area. The purpose of this book is to promote awareness of the inter-disciplinary nature of composites. New composite materials and/or new knowledge of existing composites has required "cross-fertilization" and interaction between many disciplines including: chemistry, macromolecular chemistry, surface chemistry, fiber/textile fundamentals, chemical/process engineering, mechanical design, basic materials science, analytical chemistry, instrumentation, and mechanical fabrication.

It is anticipated that the materials science trained reader will gain an appreciation for the relevant chemistry while the chemical practitioner will gain an understanding of composite properties and structure–property relationships. Designers and process trained individuals will acquire an insight into these different disciplines.

To achieve this mission the details of tensors, micro- and macro-mechanics, laminate or design equations as well as fundamental chemical concepts are mentioned by reference. The text focuses on integrating cross-functional views and experience as opposed to emphasis on any one discipline. It is written to be "user-friendly" for a diverse reader audience engaged in composites or anticipating entry into this challenging and demanding technology.

Bound Brook/Atlanta
June 1994

Louis A. Pilato
Michael J. Michno

Table of Contents

Abbreviations XIII

Units XVII

1 Introduction 1

1.1 Origin. 1
1.2 Advanced Composite Materials, Highly Specialized FRP . . . 1
1.3 ACM Industry Structure. 3
1.4 References. 8

2 Matrix Resins 9

2.1 Introduction. 9
2.2 Thermosetting Resins 10
2.2.1 Epoxy. 11
2.2.2 Phenolic Resins 18
2.2.3 Polyimides 23
2.2.4 Addition Polyimides (API). 24
2.2.4.1 Norbornene 24
2.2.4.2 Acetylene 30
2.2.4.3 Cyclobutene. 32
2.2.4.4 Bismaleimides 34
2.2.5 Cyanate Esters 39
2.3 Summary of Thermoset Resin Performance and Future Prospects 42
2.4 Thermoplastic Resins 42
2.4.1 Polyaryl Ethers 43
2.4.1.1 Nucleophilic Displacement Polymerization 43
2.4.1.2 Friedel-Crafts Polymerization 48
2.4.2 Thermoplastic Polyimides (TPI) 50
2.4.2.1 T_g Correlation. 56
2.4.3 Polyarylene Sulfide 56
2.4.4 Summary of Thermoplastic Resin Performance and Future Prospects 57
2.5 Ordered Molecules 57
2.6 Molecular Composites. 61
2.6.1 Molecular Entanglement. 62

2.6.2 Graft Copolymer/In situ Polymerization 62
2.6.3 Block Copolymer 63
2.7 Interpenetrating Networks 65
2.7.1 Semi-Interpenetrating Networks 65
2.8 Alloys or Blends 67
2.9 References 69

3 High Performance Fibers 75

3.1 Introduction 75
3.1.1 Fiber Modulus 77
3.1.2 Fiber Strength 77
3.1.3 Fiber Compressive Strength 78
3.2 Ultra High Molecular Weight Polyethylene 79
3.3 Aramid Fibers 81
3.3.1 Comparison of Different Aramid Fibers 83
3.4 Carbon Fibers 86
3.4.1 Rayon Precursor 87
3.4.2 Pitch Precursor 88
3.4.3 PAN Precursor 89
3.5 S-Glass 92
3.5.1 Comparison of S-Glass and E-Glass 93
3.6 Other Fibers 93
3.7 Summary of High Performance Fibers 94
3.8 References 94

4 Analysis/Testing 97

4.1 Introduction 97
4.2 Fiber 97
4.2.1 Surface/Structural Methods 97
4.3 Resin/Prepreg 99
4.3.1 Thermoset 99
4.3.1.1 Chromatography and Thermal Analysis Method 99
4.3.2 Thermoplastic 100
4.3.3 Blends/IPN 101
4.4 Composite 101
4.4.1 Destructive Tests 102
4.4.2 Non-Destructive Tests 103
4.4.2.1 Sound/Sonics 103
4.4.2.2 X-Ray 105
4.4.2.3 Interferometric Methods 105
4.4.2.4 Other Techniques 106
4.5 References 107

5 Composite Interphase . . . 108

5.1 Introduction . . . 108
5.2 Interphase Development . . . 108
5.2.1 Chemical/Physical Bond . . . 108
5.2.2 Surface Effects . . . 109
5.2.2.1 Fiber . . . 109
5.2.2.2 Matrix Resin . . . 110
5.3 Characteristics of the Interphase . . . 111
5.4 Interfacial Bond Strength . . . 111
5.5 Future Efforts . . . 112
5.6 References . . . 113

6 Composite Mechanical Properties . . . 114

6.1 Introduction . . . 114
6.2 Lamina Properties . . . 114
6.3 Laminate Properties . . . 118
6.4 Summary . . . 119
6.5 References . . . 119

7 Composite Structure-Property Guidelines . . . 120

7.1 Introduction . . . 120
7.2 Composite Moduli . . . 120
7.3 Composite Strength . . . 121
7.4 Edge Delamination Strength . . . 123
7.5 $\pm 45^\circ$ Tension . . . 126
7.6 Summary . . . 127
7.7 References . . . 127

8 Composite Compressive Strength . . . 128

8.1 Introduction . . . 128
8.2 Compressive Strength/Structure Property Status . . . 128
8.3 Compression Testing Methods . . . 129
8.4 Axial Compressive Strength Requirements and Properties . . . 131
8.4.1 Open-Hole-Compression Testing . . . 133
8.5 Summary . . . 134
8.6 References . . . 134

9 Damage Tolerant Composites: Post Impact Compressive Strength . . . 136

9.1 Introduction . . . 136
9.2 Compressive-Strength-After-Impact . . . 137
9.2.1 Standard Tests . . . 137

9.3 Toughened Matrix Resins 138
9.4 Interleaf Concept 139
9.5 3-D Reinforced Fiber Preforms. 140
9.6 Structure Property Trends 140
9.7 Representative Damage Tolerant Systems 141
9.8 Summary 142
9.9 References. 142

10 Thermoplastic Composites 144

10.1 Introduction. 144
10.2 High Performance Thermoplastic Matrices 144
10.3 Thermoplastic Prepreg Product Forms. 145
10.3.1 Quadrax 146
10.3.2 Electrostatic/Powder. 146
10.3.3 Commingled Yarns 146
10.4 Processing Technology. 147
10.5 Assembly Techniques 148
10.6 Polyarylether Sulfone/Carbon Fiber Composites 149
10.6.1 Processing. 150
10.6.2 Composite Laminate Properties 155
10.7 Summary 155
10.8 References. 155

11 Applications 157

11.1 Aircraft 157
11.1.1 Structural 157
11.1.2 Aircraft Interior Components 160
11.1.3 Aircraft Brakes 162
11.2 Ballistics 163
11.2.1 Components for Ballistic Composites. 164
11.2.2 Future Opportunities for Ballistic Composites 167
11.3 Space 167
11.3.1 Launch Systems 167
11.3.2 Self-Contained Space Modules 168
11.3.2.1 Commercial and Military Satellites. 168
11.3.2.2 Spacecraft 169
11.3.2.3 Space Stations. 169
11.3.2.4 Space Shuttle 170
11.3.3 Future Opportunities in Space 171
11.4 Sports/Leisure. 171
11.4.1 Golf. 172
11.4.2 Tennis Rackets 173
11.4.3 Bicycles 173

11.4.4 Skis 174
11.4.5 Fishing Rods 175
11.4.6 Archery 175
11.4.7 Other Miscellaneous Sports 175
11.4.8 Future Efforts in Sport Equipment 175
11.5 Naval Vessels 176
11.6 Tooling 177
11.6.1 Future Efforts in Tooling 179
11.7 Automotive 179
11.8 Miscellaneous Uses 181
11.8.1 Medical Uses 181
11.8.2 Industrial Machinery 181
11.8.3 Racing Cars 182
11.8.4 Non-Automotive Composite Drive Shafts 182
11.8.5 Construction 183
11.8.6 Transportation (Rail cars, trains) 183
11.8.7 Masts, Antennas, Radomes. 184
11.8.8 Musical Instruments 184
11.8.9 Non-Ballistic Helmets 185
11.8.10 Others 185
11.9 References 185

Subject Index 187

Abbreviations

ABA	Sequence Block Copolymer
ACM	Advanced Composite Materials
AE	Acoustic Emission
AIS-U	Acoustography
API	Addition Polyimides
ASTM	American Society for Testing and Materials
ATF	Advanced Tactical Fighter
ATL	Automatic Tape Laydown
ATR FTIR	Attenuated Total Reflectance Fourier Transform Infrared Spectroscopy
AUSS-V	Automatic Ultrasonic Scanning Systems
AXAF	Advanced X-Ray Astrophysics Facility
BMAC	Boeing Military Airplane Company
BMI	Bismaleimide
BT	Bismaleimide Triazine
CAD/CAM	Computer Aided Design/Computer Aided Manufacturing
CAI	Compression Strength After Impact
C–C	Carbon Carbon Composite
CF	Carbon Fiber
CFRC	Carbon Fiber Reinforced Concrete
CFRP	Carbon Fiber Reinforced Plastic
CIFV	Composite Infantry Fighting Vehicle
CPI	Condensation Polyimides
CP/MAS	Cross Polarization/Magic Angle Spinning
CRAG	Composite Research Advisory Group
CRT	Cathode Ray Tube
CTBN	Carboxy Terminated Butadiene Nitrile Rubber
CTE	Coefficient of Thermal Expansion
CV	Coefficient of Variation
CVD	Carbon Vapor Deposition
DDS	Diamino Diphenyl Sulfone
DEA	Dielectric Analysis
DMA	Dynamic Mechanical Analysis
DMF	Dimethyl Formamide
DMSO	Dimethyl Sulfoxide
DRIFT	Diffuse Reflectance Infrared Fourier Transform
DSC	Differential Scanning Calorimetry
DTA	Differential Thermal Analysis

EDS	Edge Delamination Strength
ELV	Expendable Launch Vehicles
ESPI	Electron Speckle Pattern Interferometry
FA	Failure Analysis
FAA	Federal Aviation Administration
FST	Flame-Smoke-Toxicity
FTIR	Fourier Transform Infrared Spectroscopy
GEO	Geosynchronous Earth Orbit
GPC	Gel Permeation Chromatography
GRO	Gamma Ray Observatory
HBA	*p*-Hydroxy Benzoic Acid
HDT	Heat Distortion Temperature
HERTIS	High Energy Real Time Inspection System
HNA	Hydroxy Naphthanoic Acid
HPLC	High Performance Liquid Chromatography
HST	Hubble Space Telescope
HTS	High Tensile Strength
HUMMER	High Mobility Multipurpose Wheeled Vehicle
IFSS	Interfacial Shear Strength
IITR	Illinois Institute of Technology Research Institute
ILSS	Interlaminar Shear Strength
IPN	Interpenetrating Polymer Networks
ISS	Ion Scattering Spectroscopy
LARC	Langley Research Center
LCP	Liquid Crystalline Polymers
LEO	Low Earth Orbit
MDI	Methylene Dianiline
MPD-I	Meta Phenylene Isophthalamide
Mn	Number Average Molecular Weight
MRI	Magnetic Resonance Imaging
Mw	Weight Average Molecular Weight
MW	Molecular Weight
MWD	Molecular Weight Distribution
NASA	National Aeronautical and Space Agency
NASP	National Aerospace Plane
NDT	Non Destructive Testing
NIST	National Institute of Standards and Technology
NMP	*N*-Methyl Pyrrolidone
NMR	Nuclear Magnetic Resonance
OEM	Original Equipment Manufacturer
OHC	Open-Hole Compression
OHT	Open-Hole Tension
OSU	Ohio State University

PAA	Polyamic Acid
PAE	Polyaryl Ether
PAEK	Polyaryl Ether Ketone
PAI	Polyamide Imide
PAN	Polyacrylonitrile
PAS	Positron Annihilation Spectroscopy
PASGT	Personnel Armor System-Ground Troops
PBI	Polybenzimidazole
PBO	Polybenzoxazole
PBOX	Phenylene Bisoxazoline
PBT	Polybenzothiazole
PEEK	Polyether Ether Ketone
PEI	Polyether Imide
PEK	Polyether Ketone
PEKK	Polyether Ketone Ketone
PF	Phenol Formaldehyde Resin
PI	Polyimide
PIC	Post Impact Compression
PMR	Polymerization from Monomeric Reactants
PPBP	*p*-Phenyl Biphenol
PPD-T	Poly-*p*-Phenylene Terephthalamide
PPS	Polyphenylene Sulfide
PSF	Polysulfone
RAE	Royal Aircraft Establishment
RS	Raman Spectroscopy
RTM	Resin Transfer Molding
SACMA	Suppliers of Advanced Composites Materials Association
SAMPE	Society for Advanced Material and Process Engineering
SANS	Small Angle Neutron Scattering
SAXS	Small Angle X-Ray Scattering
SEC	Size Exclusion Chromatography
SEM	Scanning Electron Microscopy
SIMS	Secondary Ion Mass Spectroscopy
SIRTF	Space Infrared Telescope Facility
SLAM	Scanning Laser Acoustic Microscope
SMC	Sheet Molding Compound
SPI	Society of Plastics Industry
SSLCP	Super Strong Liquid Crystalline Polymer
STM	Scanning Tunnelling Microscope
T_g	Glass Transition Temperature
Tm	Crystalline Melting Temperature
TBA	Torsional Braid Analysis
TEM	Transmission Electron Microscopy
TGA	Thermal Gravimetric Analysis

TICA	Torsion Pendulum [Impregnated Cloth] Analysis
TLC	Thin Layer Chromatography
TMA	Thermal Mechanical Analysis
TP	Thermoplastic
TPA	Terephthalic Acid
TPI	Thermoplastic Polyimides
TS	Thermoset
TTT	Time-Temperature-Transformation Isothermal Cure Diagram
UDRI	University of Dayton Research Institute
UHMWPE	Ultra High Molecular Weight Polyethylene
UV	Ultraviolet
V_f	Volume Fraction
WAXS	Wide Angle X-Ray Scattering
XPS	X-Ray Photoelectron Spectroscopy

Units

	U.S. Units	Factor	S.I. Units
Length	in	2.54	cm
Area	in^2	6.45	cm^2
Density	lb/in^3	2.768×10^4	kg/m^3
Yield	yd/lb	2×10^{-3}	m/g
Stress (or pressure)	lb/in^2	6.897×10^{-3}	MPa
Impact	ft/lb	1.356	J
Thermal conductivity	BTU-ft/h(ft^2)°F	1.73	W/mK
Linear thermal expansion	in/in/°F	1.8	m/m/°C
Viscosity	Poise	0.1	Pa s

1 Introduction

1.1 Origin

Advanced Composite Materials (ACM) are relatively new materials that derive their origin from the embryonic efforts of the Fiberglass Reinforced Plastics (FRP) industry that developed during the 1940s. The combination of glass fiber with a thermoset resin matrix system was instrumental in launching a commodity industry that currently provides FRP products in a diversity of market areas such as transportation, building/construction, marine, electrical/electronic and consumer products (Table 1). Many of these industrial composites are cost competitive with metals and in many instances are able to displace metals due to improved properties (low corrosion and better fatigue), lower tooling costs and ease of fabrication.

1.2 Advanced Composite Materials, Highly Specialized FRP

The small aircraft/aerospace/military segment is the highly specialized portion of the FRP market and differs from conventional FRP due to requirements for superior performance under extreme mechanical, electrical or environmental conditions. ACM can best be characterized as materials that are governed primarily by the properties of the reinforcing fibers which have high strength and high stiffness characteristics and occupy a high volume fraction of the composite. Due to their low density, unusually high axial or longitudinal specific strength and stiffness values are obtained. ACM are driven by the requirements of highly specialized market segments where, in many cases, component cost is dictated by performance. ACM are the high value, low volume segment of FRP. By the year 2000 the volume of ACM is projected to be 30,000 tonne with a corresponding value of about $3–5 billion [1]. Use of ACM is primarily identified with aircraft/aerospace and leisure or sports applications. High growth potential exists in early penetration into special industrial/commercial applications as technology transfer occurs from pioneering military applications. These specialty market segments have facilitated the replacement of steel and/or aluminum by ACM. ACM provide higher stiffness, strength and fatigue resistance as well as lower weight and lower thermal expansion when they are compared with steel or aluminum.

As one replaces metals with ACM, the design function is integrated into the total system design (Fig. 1) [2]. This is understandable since there is a diversity

Table 1. FRP composites market in 1991

Market area	Volume (1,000 ton)
Transportation	324
Construction	205
Marine	169
Corrosion resistant equipment	163
Electrical/electronic	110
Consumer products	75
Appliances/business equipment	70
Aircraft/aerospace/military	18
Other	37
	$1{,}171 \times 10^3$ ton/year

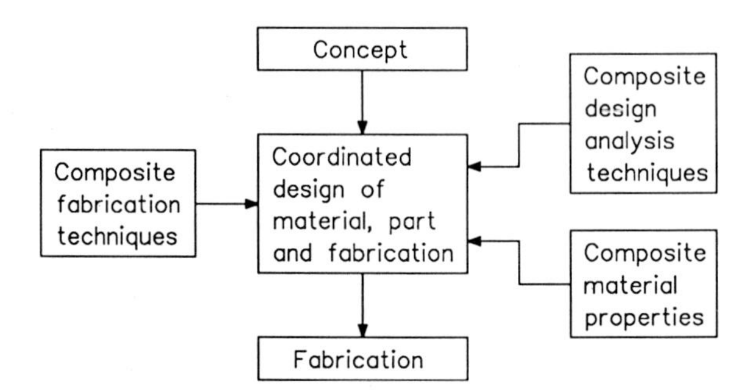

Fig. 1. Total system design

of high strength reinforcing fibers and matrix resins that can be considered in the design of the composite component. Hence, the materials designer is a member of the team that considers the composite part from inception to fabrication. A new design dimension becomes available whereby design and material considerations are contemplated simultaneously for the application rather than the customary limitation of design to available materials. Design freedom is increased by ACM. The ability to tailor properties of composites quite often makes composites the only materials capable of transforming new design concepts into reality.

Structures made of ACM help to reduce production costs by reducing the number of parts in the fabrication of the final product. In aerospace applications, parts consolidation has been demonstrated resulting in structures with fewer parts, joints, connections, and fasteners. In helicopter manufacture the total number of parts was reduced by a factor of 7 (from 11,000 to 1500) through the use of ACM. There was a corresponding reduction (90%) in the number of fasteners. The ultimate result was a safer, better performing, more efficient helicopter at a cost comparable to a helicopter produced with conventional materials.

1.3 ACM Industry Structure

The structure of the ACM industry is quite complex with reinforcing fiber and matrix resin manufacturers as the dominate raw material suppliers. Fibers and resins are combined into intermediate compounds or prepregs (pre-impregnated fibers) and fabricated into finished components. A flow diagram of various industrial participants of the ACM industry is shown in Fig. 2 [3]. There can be overlap of function with industrial participants. Some fiber manufacturers have integrated forward and produce prepreg (Amoco, Hercules, etc.). Resin producers manufacture prepreg and/or compounded ACM (ICI, Ciba Geigy, etc.). The degree of integration in the downstream fabrication or use of ACM varies significantly by end use market. Aerospace OEM's perform most of their ACM part fabrication whereas the manufacture of rocket motor cases by filament winding is often conducted by contract.

This highly specialized market segment has been responsible for an unusually large number of acquisitions and joint ventures that have occurred among several companies engaged in ACM during the early 1980s. Frequently these acquisitions have strengthened and increased the capabilities of a few large companies as major ACM suppliers. These include Amoco, Hercules, Dupont, ICI, Hoechst Celanese, British Petroleum and Ciba Geigy. However, the world events and slow economic growth may stimulate an ACM industry rationalization to further consolidate materials producers and enhance economic viabililty.

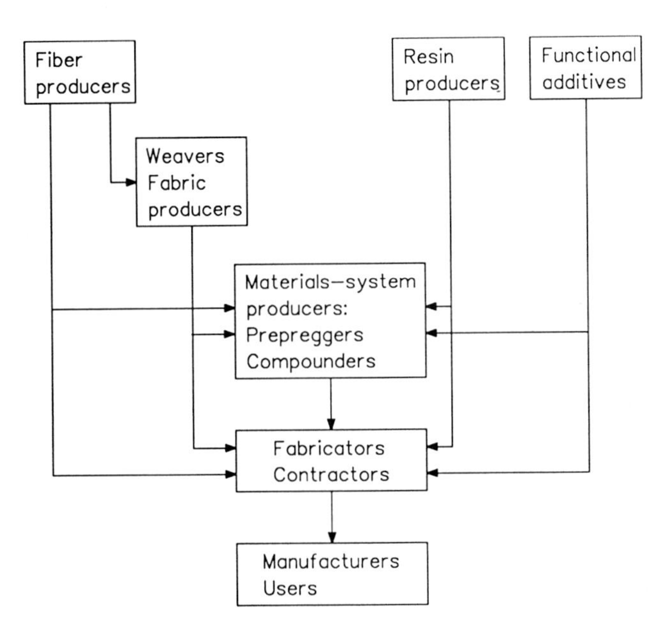

Fig. 2. Diversity of ACM industry

During the dynamic growth period of ACM from the 1950s to the 1980s major technical efforts were focused into military/aerospace markets which were heightened by the Cold War and the "race for space". Major funding by the Pentagon (U.S. Military) and the National Aeronautical and Space Agency (NASA) was the main driving force in establishing in the U.S. a preeminent position in the development of military aircraft and auxiliary defense equipment as well as a highly successful space program. These highly successful efforts are presently identified with many recent military aircraft: B-2 (Fig. 3), F-22 (Fig. 4), V-22 (Fig. 5), stealth equipment; military personnel equipment; ballistic helmet (Fig. 6), ballistic vest; CIFV (Fig. 7), light weight, portable composite bridge, electronic equipment. Synergies developed between military and commercial aircraft manufacturers resulted in a "pull through" of technology in adapting ACM in commercial aircraft. The Airbus 340 (Fig. 8) has 15% ACM while the newly planned Boeing 777 (Fig. 9) will have 13% ACM.

The other important market sector of ACM is sports/leisure equipment. Strength and stiffness/rigidity provided particularly by carbon fiber composites has revolutionized sports equipment especially tennis rackets, golf clubs, fishing rods and archery bows. Tailorability of ACM is one of the great attributes of ACM especially for sports equipment. ACM and CAD/CAM techniques have dramatically changed the sports equipment market area by achieving design flexibility in sports hardware which is safer, stronger, stiffer, lighter, more durable and better able to dampen vibrations. These newly developed high technology sports equipment clearly rivals all the early models that were composed of metal or wood.

Aerospace and Sporting Goods market sectors continue to provide technical and economic challenges where superior performance is able to accommodate

Fig. 3. B-2

Fig. 4. F-22

Fig. 5. V-22

the prevailing cost of expensive raw materials. The application of the same ACM technology to other market areas has been slow and obviously not as dynamic as the previously mentioned sectors. These same technical and economical parameters are not entirely applicable to more mature large volume industries such as automotive, construction and mass transportation where premium properties of ACM could possibly make an impact. In fact ACM can offer significant weight and parts consolidation advantages for these markets. Yet ACM are perceived as overqualified especially for current automotive needs. Economical E-glass provides satisfactory strength as compared to expensive carbon fiber. Furthermore the automotive industry is predominately highly

Fig. 6. Ballistic Helmet

Fig. 7. CIFV

automated with high production rates and a highly competitive pricing structure of raw materials. If the ACM component cannot be manufactured with high quality, high productivity and competitive cost, the full potential of automotive ACM will not be realized regardless how exceptional the component, the design or their combined benefit becomes.

In spite of these somber statements, ACM drive shafts via filament winding have been introduced in some General Motors vehicles. As the Clean Air Act (U.S.) and its amendments are implemented in the 1990s, automotive fuels are undergoing changes. Alcohol/ether fuel blends are being considered to reduce

Fig. 8. Airbus A340, Courtesy of Airbus Industrie

Fig. 9. Boeing 777

pollution caused by vehicles and improve air quality. Comparison studies between normal gasoline, alcohol/ether fuel blends and natural gas suggest that natural gas is the cleanest burning fuel. U.S. vehicle manufacturers are currently testing natural gas as fuel in several test vehicles (autos, trucks, buses). Fuel storage of natural gas will require a high pressure tank (Fig. 10) with sufficient fuel capacity to provide a reasonable driving range. ACM via filament winding are ideal for these high pressure tanks. Thus special or niche markets exist for ACM in these mature market areas. These are described in more detail in Chapter 11 on applications.

In the late 1980s, the collapse of the Soviet Republic and the end of hostilities between the Western nations and the Soviet dominated nations (Cold War) has resulted in reduced military spending and cancellation of some defense programs. This has prompted many companies engaged in ACM to reassess this

Fig. 10. High Pressure Tank

highly specialized and technology intensive area. Since the major focus of ACM was driven by defense/space related market areas with minimal opportunity in mature, large markets, some companies have divested their ACM activities (Phillips, BASF) while others have formed joint ventures (Dow-United Technology) or strategic alliances (Dupont-Hercules) in selected ACM areas. Over capacity in fiber manufacture and the broad spectrum of matrix resins suggests that a further consolidation phase will occur among raw materials manufacturers.

Besides military/aerospace, civilian aircraft and sports equipment markets, specialized or niche markets will be exploited by ACM. Continued development of ACM is inevitable in many different industries based on past technological achievements that have generated new materials designed to fulfill multiple functions through parts consolidation. Future efforts in ACM will integrate the concept of concurrent engineering whereby design, processing and economics are simultaneously considered to manufacture reliable, reproducible ACM with quality and multifunctionality built in. Concurrent engineering will be coupled with newly emerging technology related to smart systems wherein processing is controlled through use of sensors and control systems. In some selected applications smart systems will evolve into smart/intelligent structures which would signal flaws or weaknesses within the ACM component prior to catastrophic failure.

1.4 References

1. J. McDermott, Advanced Composites, 1991 Bluebook, p. 7.
2. M. Salkind, Proc. ICCM, Am. Ind. Met. Eng. New York 1976, Vol. 2, p. 5.
3. Modern Plastics, p. 72, July 1989.

2 Matrix Resins

2.1 Introduction

Advanced composite materials consist of a high strength reinforcing agent or fiber combined with a high performance matrix resin. The high performance matrix resin is the continuous phase in which the reinforcing agent or fiber is contained. The matrix resin provides uniform load distribution to the fiber and protects or safeguards the composite surface against abrasion or environmental corrosion, either of which can initiate fracture. Adhesion between the matrix resin and the reinforcing agent is necessary to allow uniform load distribution to occur between the two dissimilar phases. However in some special instances, particularly for ballistic components containing either aramid fiber or high performance polyethylene fiber in a matrix resin, minimal adhesion between fiber and matrix resin is preferred.

The high performance matrix resin should possess a modulus of at least 3 GPa for optimum strength and sufficient shear modulus to prevent buckling of fiber under compression load.

A further important feature of the matrix resin is that it should absorb energy and reduce stress concentrations by providing fracture toughness or ductility to maximize damage tolerance and long term durability. It must also provide hot/wet performance. Susceptibility of the matrix resin to water and the corresponding reduction of hot/wet performance in a hostile environment is a concern. Continued cycling from dry to wet environments may cause a slight increase in polymer volume or swelling. The reversible swelling/shrinkage of resin matrix under these conditions may cause microcracking. Thus the matrix resin is expected to provide optimum composite characteristics within a prescribed temperature range in spite of these aforementioned problems. The ultimate thermomechanical performance characteristics of the composite are governed by the heat resistance of the matrix resin.

A variety of thermosetting and thermoplastic resin systems have been utilized as matrix resins. A comparison of advantages and disadvantages of these contrasting resin systems as they apply to resin, prepreg or composite is shown in Table 1.

In selecting a matrix resin the differences between a thermoset or thermoplastic resin must be assessed. Each of these resin families possesses strengths and weaknesses that must be evaluated. The ultimate selection of the matrix resin encompasses the resin, its transformation into prepreg by combining with reinforcing fiber and finally by considering composite processing, fabrication and properties.

Table 1. Composites based on thermoset or thermoplastic resins

Component	Property	Thermoset	Thermoplastic
Resin	Formulation	Complex	Simple
	Melt Viscosity	Very Low	High
	Fiber Impregnation	Easy	Difficult
	Cost	Low to Medium	Low to High
Prepreg	Tack, Drape	Good	None to Fair
	Stability (Shelf Life/Out Time)	Poor	Excellent
	Quality Assurance	Fair	Excellent
Composite	Processing Cycle	Long	Short to Long
	Processing (Time/Temp/Pressure)	Low to Moderate	High
	Fabrication Cost	High	Low (Potentially)
	Structural Properties	Fair to Good	Fair to Good
	Environmental & Solvent Resistance	Good to Excellent	Poor to Good
	Interlaminar Fracture Toughness	Low	High
	Damage Tolerance	Poor to Excellent	Fair to Good
	Data Base	Large	Small

2.2 Thermosetting Resins

Thermosetting resins are low molecular weight telechelic oligomers with low to medium viscosity but require a catalyst and/or hardening agent or elevated temperature cure conditions (> 100 °C). Frequently post curing is also necessary for maximum mechanical properties. The resulting rigid resin system is converted into a highly crosslinked, network macromolecule exhibiting excellent environmental and solvent resistance, high mechanical strength/stiffness with minimal toughness.

A series of different events occurs as the reactive oligomer(s) is cured. During time and temperature the reactive molten liquid is transformed through various intermediate stages from liquid to gel and ultimately to a highly cured, crosslinked or vitrified state. Gillham [1] has described these different events that a reactive thermosetting system undergoes, particularly epoxy resins, with a time-temperature-transformation (TTT) isothermal cure diagram. Little reaction occurs after vitrification due to immobility of the reacting groups. When vitrification inhibits further reaction, the T_g of the system will be the isothermal cure temperature. If the isothermal cure temperature is below the maximum T_g of the thermoset system, vitrification can occur prior to maximum conversion.

When this occurs, postcuring above maximum T_g is necessary for the development of optimum properties (this is necessary for phenolics or acetylene terminated oligomers). For some highly crosslinked systems the maximum T_g can exceed resin thermal stability, and in those instances maximum conversion of the network forming reactions is not possible. Usually the curing characteristics of thermosetting resin systems are so designed (TTT diagram, rheological data or dynamic mechanical analysis) that the T_g and cure temperature are predetermined for optimum processing and composite temperature requirements. Additionally one must also consider whether volatile by-products are released during the crosslinking process (Phenolics, Condensation Polyimides).

The most prominent thermosetting resin systems that have been utilized in high performance advanced composites are epoxies, phenolics, bismaleimides, polyimides, and cyanate ester resin systems. Unsaturated polyester and vinyl ester thermosetting systems are not described due to low to intermediate T_g properties of the crosslinked product. Although these earlier mentioned thermoset resin systems exhibit excellent solvent resistance with high heat resistance, some systems particularly epoxies and BMI resins are susceptible to hot/wet temperature transformations and result in lower (reduced) T_gs for the crosslinked material.

2.2.1 Epoxy

Epoxy resins can vary from difunctional to polyfunctional epoxides as monomers or prepolymers that react with curing agents to yield high performance thermosetting resin systems [2–5]. The attractive performance characteristics of the epoxy family of resins are exemplified by a combination of high rigidity, solvent resistance and elevated temperature behavior.

Difunctional Epoxy Resins

The most widely used difunctional epoxy resin is the diglycidyl ether of bisphenol A (I):

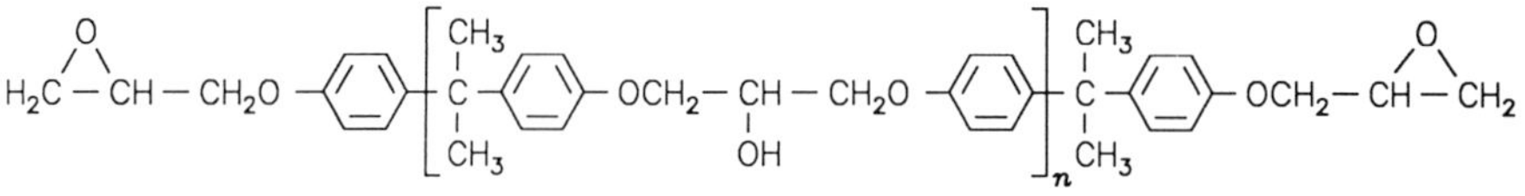

$$\overset{O}{\overbrace{H_2C-CH}}-CH_2O-C_6H_4-\left[\underset{CH_3}{\overset{CH_3}{C}}-C_6H_4-OCH_2-\underset{OH}{CH}-CH_2O-C_6H_4\right]_n-\underset{CH_3}{\overset{CH_3}{C}}-C_6H_4-OCH_2-\overset{O}{\overbrace{CH-CH_2}}$$

I

Bisphenol A epoxy resin has gained wide acceptance in a diversity of applications and is available in a range of molecular weights ($n \geq 0$). The value of "n" can vary from 0.2 to greater than 12 with Mn from about 400 to 4200.

Multifunctional Epoxy Resins

Epoxidized novolaks (II) based on the reaction of epichlorohydrin with a phenolic resin (novolak) possess multiepoxy functionality of at least 2 to more than 5 epoxy groups per molecule. Tightly crosslinked, highly cured systems with improved thermal and chemical resistance when compared to difunctional bisphenol A resins are obtained. The T_g of the epoxy novolak is also higher than the T_g of bisphenol A epoxy resin.

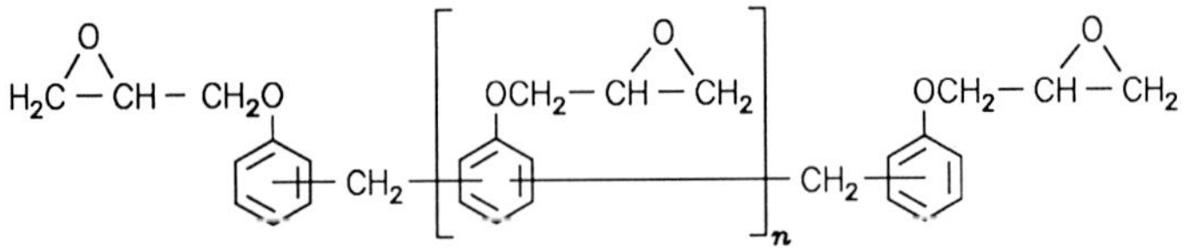

II

Specialty polyfunctional epoxy resins, particularly attractive for high temperature aerospace applications, consist of aromatic and heterocyclic glycidyl amine resins. The most popular resin within this class of specialty resins is tetraglycidyl methylene dianiline (TGMDA) III.

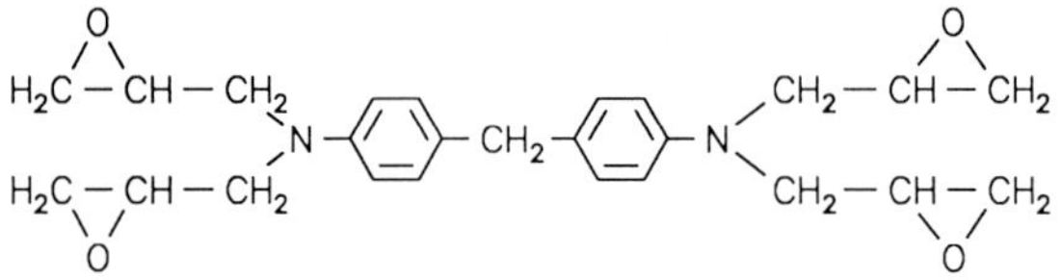

III

More recently [6] newer aromatic/heterocyclic glycidyl amine resins with improved hot/wet temperature characteristics without a corresponding reduction in T_g have been reported (IV, V):

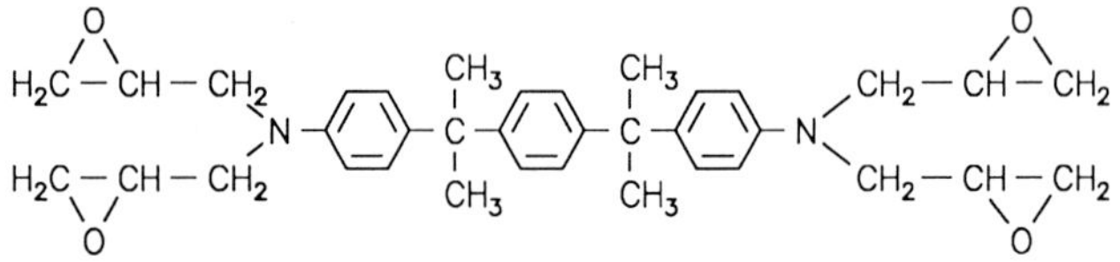

IV

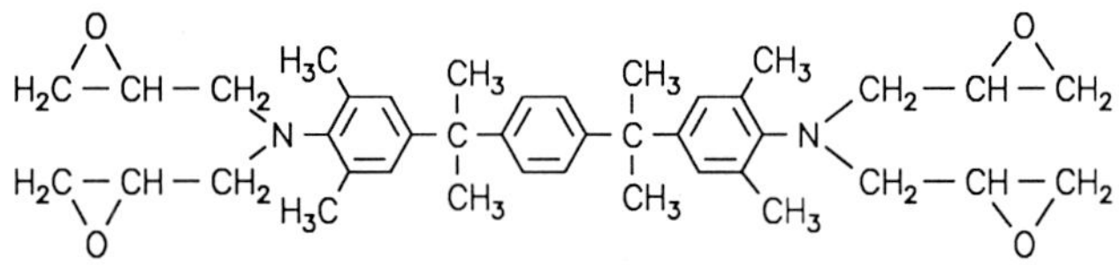

V

Besides these above mentioned epoxy resins, there are other speciality multifunctional resins which may be formulated with TGMDA and/or bisphenol A epoxy resin. These include VI, VII, VIII, IX, X:

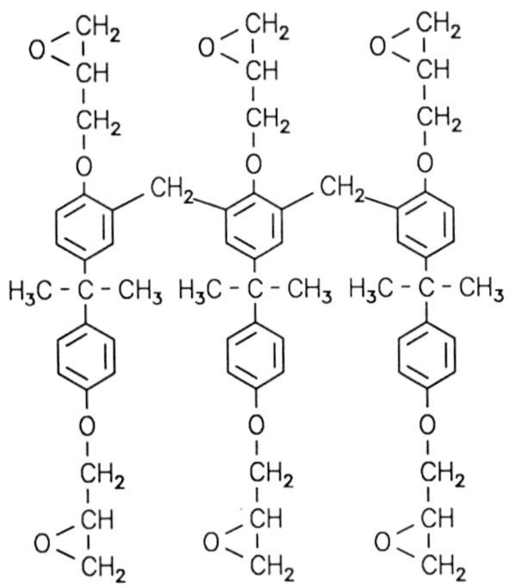

VI

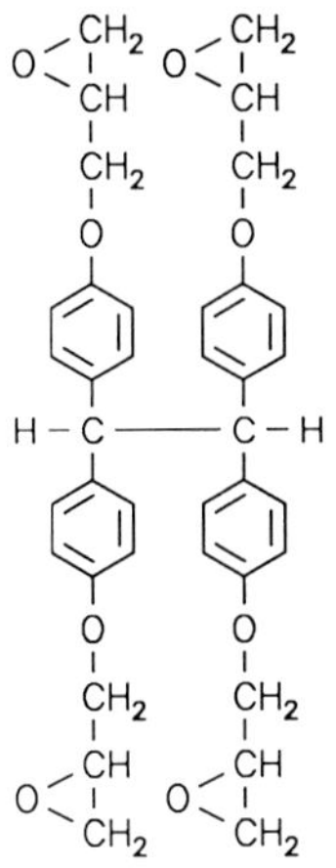

VII

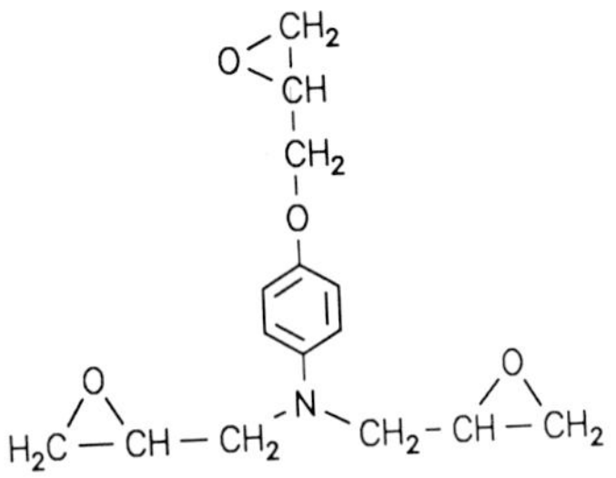

VIII

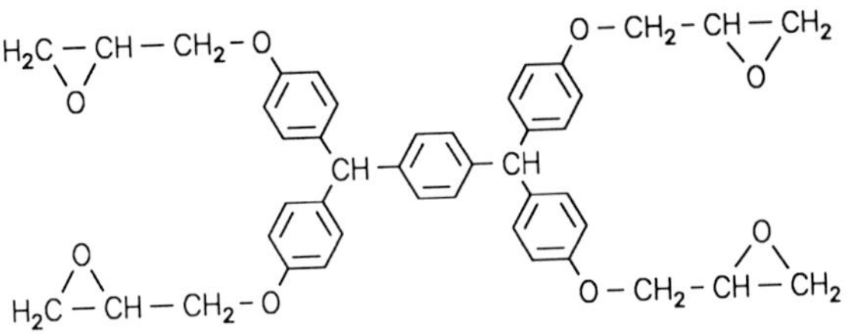

IX

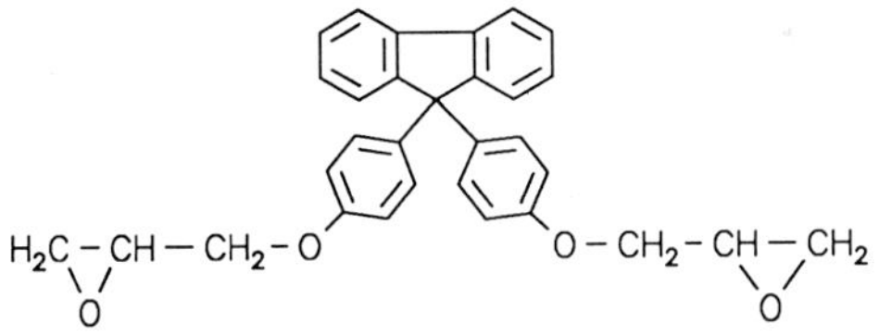

X

Structure property correlations provide some guidance in the selection of appropriate epoxy resins for the desired end use application. Resins can vary from pure monomer, telechelic oligomers or a mixture of 2 to 3 different types of epoxy resins. e.g., Narmco 5208 contains mainly III and some VI as epoxy resins.

Curing Agents

A key feature of epoxy resins is that the epoxide group reacts with a large number of molecules to form high performance thermoset resin networks without the evolution of by-products. Curing agents can be classified as either co-reactants and become incorporated within the macromolecular structure or those which promote crosslinking catalytically [2–8].

The co-reactants that participate directly in the crosslinked system can be basic or acidic curing agents. The basic curing agents include primary, secondary amines, polyaminoamides, while the acidic agents can be polyphenols, polymeric thiols and to some extent anhydrides.

Catalytic curing agents accelerate the crosslinking reaction at lower temperatures. Tertiary amines and BF_3 complexes are examples of catalytic curing agents.

Co-reactive curing agents are polyfunctional reagents that are used in nearly stoichiometric amounts with epoxy resin. The most frequently utilized polyamines can be aliphatic, aromatic, cycloaliphatic, or heterocyclic. Depending on the relative basicity of the polyamine, epoxy crosslinking reactions can occur readily at room temperature (aliphatic primary polyamine) or require elevated temperature conditions with aromatic amines. The same aromatic amine-epoxy reaction system can be accelerated through the use of e.g. a BF_3 complex as catalyst. Accelerators are also used in poly phenol-epoxy or anhydride-epoxy resin combinations.

Several factors are considered in the selection of a particular epoxy resin as a matrix resin. These include the type of curing agent and accelerator (if needed), molar ratio of resin to curing agent, crosslink density and the degree of cure occurring during processing. The final criteria in the selection process is the maximum heat resistance or T_g, of the epoxy resin as well as its hot/wet resistance. Thus epoxy resin structure, curing agent and reaction conditions dictate the resin T_g, moisture absorption, and performance of the epoxy resin in a hot/wet environment. As will be seen later, an epoxy resin with a moderately high T_g is effected in a hot/wet environment and results in a lower T_g due to the plasticizing effect of water on the resin.

Cured epoxy resins based on Bisphenol A epoxy resin with various amine curing agents exhibit T_g values between 110–150 °C (Table 2).

Higher T_g epoxy resins are obtained when multifunctional aromatic epoxies are used with an aromatic diamine, 4,4′ diamino diphenyl sulfone [DDS] (Table 3).

It can be noted that when the functionality of II was increased from 2.2 to 3.6 the T_g of the cured resin similarly increased to 235 °C. Thus T_g of the resulting cured epoxy resin can be as high as 279 °C when multifunctional aromatic glycidyl amine resins are reacted with DDS. However these aromatic glycidyl

Table 2. Effects of amine curing agent on T_g[a]

Curing agent	T_g (°C)
Diethylene Triamine	112
2,2′-Diamino m-xylene	135
P-Diamino diphenyl sulfone	150

[a] Bis Phenol A epoxy resin

Table 3. High T_g epoxy resins[a]

Epoxy resin	T_g (°C)
(II) 2.2 functionality	198
(II) 3.6 functionality	235
(III)	262
(IV)	249
(V)	258
(X)	279, 260[b]

[a] All DDS curing agent

[b]

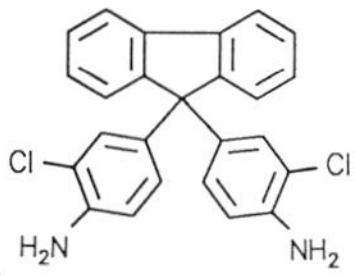

amine resins III, IV, V are moisture sensitive and absorb water with a corresponding reduction in T_g in a hot/wet environment [9]. Epoxy III or even combined with VI (composition is known as Narmco 5208) (see Table 3) and cured with DDS absorbs between 4 and 5 percent water with a corresponding reduction in T_g (< 200 °C).

Delvigs [10] has shown that aromatic diamine hardeners besides DDS can be used to cure III and lead to acceptable high heat resistant epoxy materials (T_g) with improved hot/wet resistance. Newer epoxy resins such as IV, V, or X cured with a closely related aromatic diamine XI:

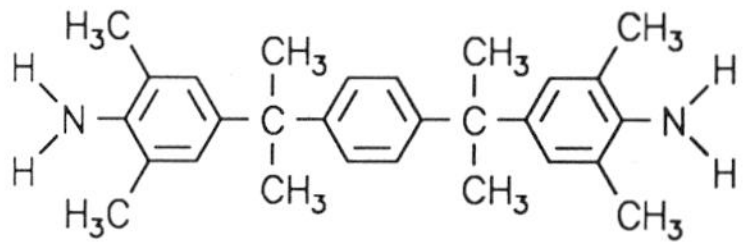

XI

have shown improved properties in a hot/wet environment [9] when these systems are compared with NARMCO 5208 and/or other TGMDA compositions with DDS (Table 4).

As it was mentioned earlier, epoxy resins are quite brittle and exhibit poor impact. Different cross link densities or intervals between cross links were examined by Fisher and Schmid [11] in epoxy novolak resins with 2.2 and 3.6 functionality. Differences in T_g and fracture energy were observed with a higher fracture energy resulting from the lower T_g epoxy novolak resin and attributable to greater mass (distance) between crosslinks.

For high temperature epoxy resins (T_g > 175 °C), various methods have been examined to improve fracture energy of the resulting cured epoxy resins. The introduction of a second phase such as reactive rubber or functional elastomers

Table 4. Mechanical properties of thermoset resins

Resin	T_g (°C)	F.S. (MPa)	F.M. (GPa)	Elong. (%)	Fracture energy (J/m^2)	Moisture gain (%)
Epoxy[a]						
I	150	114	2.9	4.4	~ 250	—
III	262	138	3.9	5.0	60–150	5.7
IV	249	117	3.9	3.7	—	3.6
X	279	124	3.3	4.7	—	2.8
III + PEI (30%)	215, 265	—	3.5	3.7	488	1.3
Phenolic						
Phenol/Formaldehyde	127 (HDT)[b]	80–100	5–7	1–1.5	—	0.1–0.2
60/40 1,3 PBOX	159	190	5.1	1–1.8	190	—
Polymide						
CPI	330	86 (TS)[c]	3.1	7.5	—	0.3
PMR-15	340	176	4.0	1.1	90–280	—
Thermid 600 (Nat'l Starch)	320	147	4.6	1.5	—	1.2
BMI						
Compimide 796 (Shell)	> 300	76	4.6	1.7	63	4.3
796/TM 123	260	132	3.9	3.5	440	3.0
796/TM 123 + 20% Polyhydantoin	294	115	3.6	3.1	454	—
Cyanate						
XXIV (HiTek)	230 (HDT)[b]	155	3.1	7	—	< 2.0
XXVI (Dow)	250–265	124	3.4	2.7	180	< 1.5

[a] All DDS Amine
[b] Heat Distortion Temp °C @ 1.82 MPa
[c] Tensile Strength

[12] has only been marginal in increasing fracture energy with a corresponding reduction in mechanical properties and T_g.

A novel approach by McGrath and coworkers [13] is to combine the most desirable features of a thermoplastic resin segment within the epoxy resin to form a multicomponent resin. Functionally terminated polysulfone XII with amine end groups (Mn of 13000 to 21000) was introduced into a multifunctional epoxy resin cured with DDS.

$$H_2N-R-O-R'\left[O-R-O-R'\right]O-R-NH_2$$

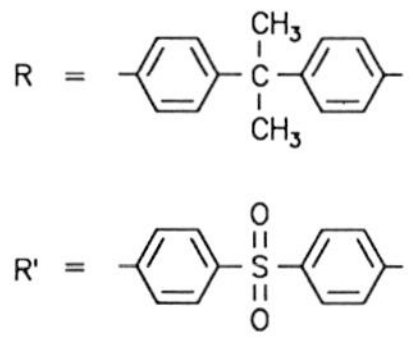

R = –C₆H₄–C(CH₃)₂–C₆H₄–

R' = –C₆H₄–SO₂–C₆H₄–

XII

The strategy devised by McGrath is based on the concept of microphase separation. The co-reacting thermoplastic modifier is initially soluble in the epoxy resin but phase separates into discrete particles during the crosslinking process. It is proposed that the particles remain ductile and "yield" during fracture with an attendant absorption of energy (Table 5).

With the increased (30% PSF) thermoplastic modifier a lower transition temperature of 180 °C for the PSF phase is observed along with a slight reduction in T_g to 191 °C for the epoxy cured resin. However fracture toughness (G_{IC}) increased five fold with a noticeable reduction in water sensitivity. Through the use of these high modulus modifiers, there is no substantial reduction in modulus of the cross linked epoxy resin as is customarily observed with rubber or elastomer modified epoxy systems.

Somewhat similar studies by Michno [14] and Kohli and Fisher [15] have involved the use of ether sulfone amine terminated materials as curing agents for

Table 5. Improved toughness of epoxy by amine thermoplastic modifier

Sample	T_g (°C)	Weight percent Water uptake[a]	Fracture	
			K_{IC} $N/m^{3/2}$	G_{IC} J/m^2
Control Resin + DDS	194	4.47	0.58	103
+ 15% (21KPSF)	193	3.92	1.06	318
+ 30% (21KPSF)	180–191	3.16	1.32	513

[a] Sample exposed to 48 hours boiling water

various epoxy resins including multifunctional epoxies (II, III, IV and V). Both research groups also introduced a thermoplastic resin in the epoxy resin/ether sulfone diamine system.

High performance thermoplastic resins such as polysulfone, polyetherimide, and polyhydantoin have been introduced into multi-functional epoxy resins particularly IV and cured with amine XI to yield toughened high performance epoxy resins [16] with improved damage tolerance without compromising the strength or hot/wet performance characteristics of the cured epoxy resin. The highest degree of toughness is provided by the polyhydantoin resin (XIII). Two phase systems or semi-IPN compositions are formed with crosslinked epoxy resin being the continuous phase and the thermoplastic resin dispersed in the continuous phase.

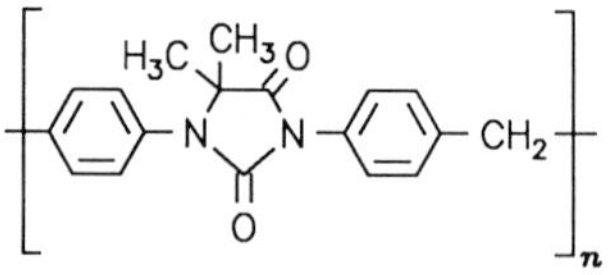

XIII

Similarly Alman [17] and Recker [18] have reported the use of selected high performance thermoplastic resins with multi-functional epoxy resins to yield tough epoxy matrix systems as semi-IPNs.

Bucknall and Gilbert [19] demonstrated that a tough epoxy resin system can be obtained by dissolving polyetherimide in TGMDA (III) followed by the addition of DDS curing agent. The PEI formed a separate phase within the cured composition. The toughened epoxy exhibited a linear increase in K_{IC} with increased PEI content with an accompanying partial reduction in modulus (Table 4).

It is likely these latter systems with a ductile, thermoplastic resin phase form semi-interpenetrating networks. In a latter section describing IPN systems, it will be shown how the IPN technique has been effective in improving the toughness of brittle resins without adversely affecting thermal or mechanical properties of the resin system.

2.2.2 Phenolic Resins

Phenol Formaldehyde (PF) resins are one of the oldest commercial polymeric resin systems and are identified with the beginning of the Plastics Industry. Their origin can be traced to the pioneering work of Leo F. Baekeland in the early 1900s [20, 21, 22]. In spite of their commodity status and severe competition arising from thermoplastic resins, PF resins continue to be utilized in many original applications that were initially developed by Baekeland. By virtue of its outstanding *flame* resistance, low *smoke* generation, and low *toxicity* emission on combustion (FST), PF are undergoing a resurgence of renewed interest in advanced composites particularly those related to aircraft interior and cargo liners. Besides use in aircraft due to FST characteristics, PF resin and modified PF resins are the resins of choice in high performance ballistic composite components that incorporate S-2 Glass or Kevlar aramid fibers.

Resin Preparation

PF resins are conveniently prepared by either an acid catalyzed reaction of phenol with less than a molar ratio of formaldehyde (< 0.85) to form a solid, low melting thermoplastic material, known as a novolak, or by a base catalyzed reaction of phenol with formaldehyde to form a resole resin. The amount of formaldehyde may vary from less than a molar ratio to more than 3 moles of formaldehyde to phenol (Eq. 1).

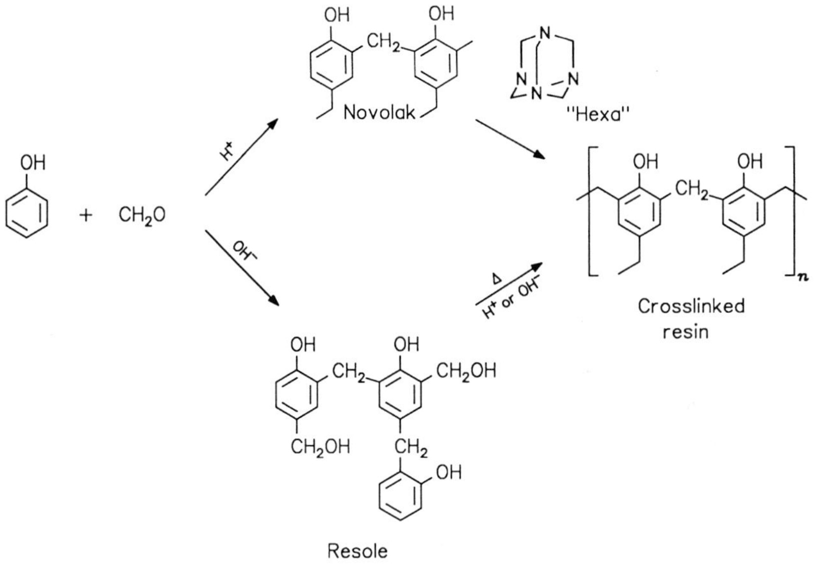

OH + CH_2O
H^+ → Novolak + "Hexa" → Crosslinked resin
OH^- → Resole → Δ, H^+ or OH^- → Crosslinked resin

Equation 1

In many cases resoles can be liquid, low to medium MW resins which are heat curable or cured by acid or base into highly crosslinked phenolic resins.

Novolaks

Novolaks require a source of formaldehyde for transformation into a cross-linked resin. This source is predominately hexamethylenetetramine. The novolak/hexa combination is commonly referred to as a "two-step" resin and is the main system that is utilized in phenolic molding materials. These latter materials are high fiber, filler and additive containing thermoset products that are compression or injection molded in many diverse electrical and automotive applications.

Water is formed as a by-product during the crosslinking of phenolic resins (novolak or resole), whereas water and ammonia are by-products during novolak curing with hexa. The cured phenolic resin system characteristics are high heat and chemical resistance, good dimensional and thermal stability with excellent surface hardness. These resin property characteristics improve with level of cure and "post-cured" compositions exhibit the best properties. Phenolics possess the best fire/smoke/toxicity (FST) characteristics of all large volume commercial resins and provide this unusual performance for a surprisingly low price.

Phenolic molding material is the system of choice for the composite motor for the Polimotor engine, a light weight engine weighing 80 kg (versus a metal based engine weighing 135 kg) with a 2.3 liter, 4 cylinder dual overhead cam 16 valve engine. The composite engine and transmission stator/reactor are a few examples of high performance automotive applications requiring advanced composite materials.

The highly crosslinked structure which provides the phenolic resin with excellent rigidity and outstanding creep also results in an inherently brittle matrix. Hence components fabricated with PF must be designed to avoid cracks from initiating and where vibrations are known to occur, fatigue stresses must be measured and compared with PF fatigue strength data.

Computer modelling of phenolic molding materials' curing behavior provides optimum phenolic postcure time-temperature cycles to obtain the desired T_g within a shorter period of time with accompanying reduced energy costs [23]. Molding material systems are available with T_gs as high as 350 °C for automotive and other high performance applications. T_gs in excess of 315 °C are becoming an increasingly common T_g requirement [24] in most metal replacement automotive and electrical applications.

Resoles

Resole resins which are liquid, low to medium viscosity resins, and usually a solution in alcohol or an aqueous dispersion [25], provide a high degree of formulation flexibility. Several curing options are available in resole crosslinking. Resoles can be cured thermally with no catalyst contamination. Thermally cured resoles are the preferred resin systems for treating glass, aramid and carbon fibers to obtain attractive FST composites for such applications as Nomex or glass honeycomb compositions, aircraft interior panels, aircraft flooring and partitions. Woven fabrics are impregnated by solvent or dispersion PF and on thermal removal of volatiles, the PF coating advances in MW to a partly crosslinked material. This latter material is commonly referred to as "B Stage" material. Depending on level of resin advancement and solvent removal, the prepregged material can exhibit tack and drape or be rigid and tack-free.

Changes in flexural, compressive, tensile strength, and modulus of phenolic glass fiber laminate (Fig. 1) for temperatures to 500 °C show that these phenolic laminates have excellent retention of properties at elevated temperatures [26].

Selected inorganic species such as Boron, Tungsten, or Zirconium compounds can be introduced [27] and co-react with the PF resin. Upon curing, these inorganic materials provide improved thermooxidative stability and increased char content (see carbon–carbon composites).

Resoles can also be catalyzed by either acid or base conditions. Newly developed resole resins catalyzed by proprietary acid catalyst are available [28] for use in RTM or pultrusion processes for FST products.

The main use of base catalyzed resoles is for sheet molding compound (SMC), recently commercialized by Occidental Chemical Corp. It is also claimed that the base catalyzed resoles are useful in pultrusion and continuous laminating processes [29].

An unusual co-reaction of phenolic resins with phenylene bisoxazolines has been reported [30]. The reaction involves a triaryl phosphine catalyzed reaction of bisoxazoline (XIV) (1,3PBOX) with a phenol free novolak resin to an ether amide copolymer [30] (Eq. 2). Mechanical properties of 40:60 composition of bisoxazoline/PF on Glass fiber compared quite favorably with two commercial

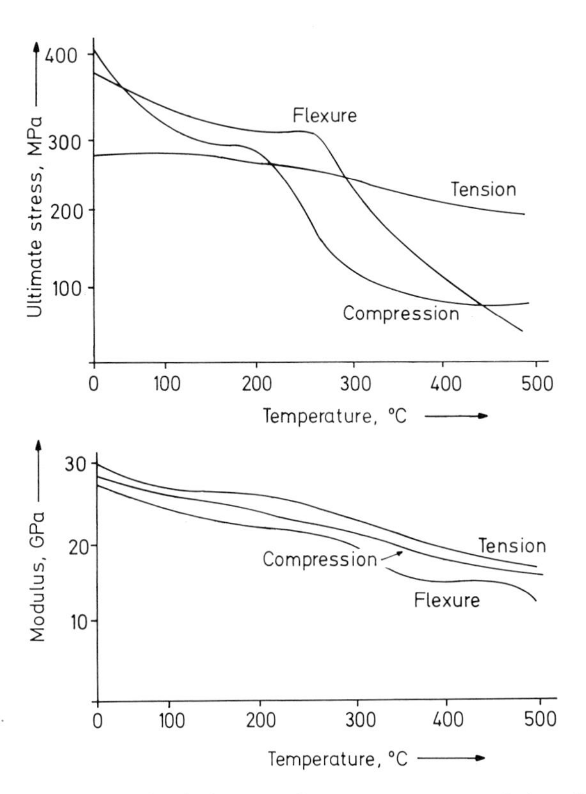

Fig. 1. Mechanical properties vs temperature of phenolic glass fiber laminates

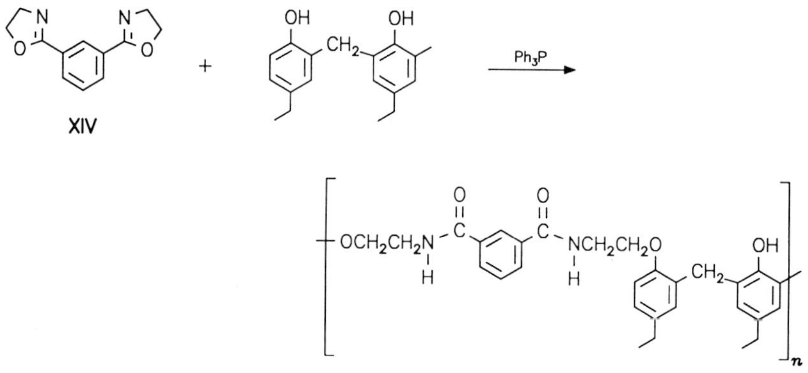

XIV

Equation 2

resin systems of epoxy/glass and PF/glass. Properties of neat phenolic resin and 60/40 1,3PBOX modified phenolic resin are recorded in Table 4.

These fiber reinforced copolymers of 1,3PBOX/PF are claimed to exhibit low smoke and low heat release (OSU values), and thus satisfactory for aircraft interior. Low warpage, low shrinkage is observed and is attributable to ring opening reaction of the bisoxazoline by the phenol moiety with no volatiles being evolved. Conditions are reported for using 1,3PBOX/PF system in a hot melt RTM application with an undisclosed reactive diluent.

It is not clear whether these copolymers with high amounts of bisoxazoline (40%) are comparable or inferior to typical phenolic systems in long term heat resistance properties in view of the aliphatic nature of the ether portion of the copolymer.

Phenolic Resin Decomposition/Char

The rigid three dimensionally crosslinked phenolic structure does not melt or soften and resists thermal stress. At elevated temperatures some decomposition occurs (300 to 600 °C) and facilitates the volatilization of low MW species. Temperatures above 600 °C result in some fragmentation and recombination or ring closure into highly condensed polycyclic systems approaching graphite-like structures. Phenolics produce little or no flammable fuel when heated by a free flame. An insulating char is produced that requires extremely high temperature for oxidation and/or an oxygen rich environment. A better description of this proposed pyrolysis of phenolics is suggested by a high resolution ^{13}C cross polarization Magic Angle spinning Solid-State NMR Spectroscopy study by Amram and Laval [31]. NMR analyses indicated that laser pyrolysis with energies from 3 000 J to 30 000 J resulted in increased graphitization with 10 condensed rings with an average cluster diameter of 9 Å for a lower energy pyrolysis and 23 condensed rings with an average cluster diameter of 14 Å for

30 000 J. The assembly of a condensed ring structure of the phenolic resin during pyrolysis is dependent on energy and time.

Carbon–Carbon Composites

The significance of thermally treating PF into high char residue has been exploited commercially by the development of carbon–carbon composites (C–C).

Carbon–carbon composites can be produced by either a liquid resin (resole or pitch) impregnation or chemical vapor deposition. The impregnation method involves the treatment of carbon fibers with a resin followed by pyrolysis into high char or carbon–carbon composite. The impregnation–pyrolysis step is repeated several times until the proper density is obtained. Phenolic resole resin is chosen because of high char yield on pyrolysis. The chemical vapor deposition (CVD) technique deposits carbon atoms within the porous carbon fiber structure. The CVD method applies a surface coating resulting in a highly anisotropic carbon matrix.

In C–C composites a bond of intermediate strength is desired between the fibers and the matrix. Usually untreated fibers are often preferred for intermediate strength of fiber/matrix.

The importance of C–C composites lies in their ability to survive beyond 2200 °C. They are light weight materials and capable of being cycled from a subzero temperature to 1500 °C in seconds with utility in high abrasion and high temperature applications such as aircraft brakes, heat shields for reentry space vehicles, engine turbines and military/aerospace market areas.

2.2.3 Polyimides

The imide structural unit within a polymeric matrix resin system has gained wide acceptance in premium high performance composites. Polyimides are the preferred resin system when unusually high temperature resistance is required i.e. > 300 °C. Polyimides are thermooxidatively and dimensionally stable and retain a high degree of mechanical strength at temperatures beyond the degradation of many polymers. Many types [32–35] of polyimide resins have been developed and consist of condensation polyimides (CPI) addition polyimides (API) or thermoplastic polyimides (TPI). Within the family of API type materials, a modified or unsaturated imide, the bismaleimide (BMI) functionality is described. The thermoplastic polyimide (TPI) family will be discussed in the thermoplastic resin section.

Condensation Polyimides (CPI)

The classical method for the preparation of CPI is the condensation of a tetracarboxylic acid dianhydride or diester diacid with an aromatic diamine (Eq. 3).

Dianhydride or Diester diacid $+ H_2N-R-NH_2 \xrightarrow[\text{NMP, DMF or DMAc}]{\text{Aprotic solvent}}$ Polyamic acid $\xrightarrow{Ac_2O/\text{Amine} \quad \Delta\ (150-300\,°C)}$ Polyimide

Equation 3

The intermediate, polyamic acid, is chemically (acetic anhydride and amine) or thermally treated to remove water and facilitate ring closure. The resulting insoluble and nearly intractable polyimide can be fabricated by utilizing the soluble polyamic acid (PAA) intermediate. Although the PAA is processable, it is hydrolytically unstable and thermal dehydration/devolatilization must be carefully controlled to avoid blistering due to removal of high boiling solvents and by-product, water. With a lesser reacting monomer such as diester diacid, the PAA can be obtained in higher concentrations in the aprotic solvent. By end capping PAA with dicarboxylated monoanhydrides, high solids, low viscosity solutions of PAA with high MW can be obtained [36].

The absence of water is critical. Besides hydrolyzing anhydride monomer (if used) the intermediate PAA is moisture sensitive and is cleaved with a corresponding reduction in MW of PAA. Further, an ambient temperature range of 10 to 35 °C is desirable. If higher temperatures are used, premature imidization occurs followed by the precipitation of low MW insoluble product.

Polyimide can also be prepared by the reaction of dianhydride with a diisocyanate or by an aromatic nucleophilic displacement technique. Both of these methods lead to thermoplastic polyimides and will be described in the thermoplastic resin section.

2.2.4 Addition Polyimides (API)

2.2.4.1 Norbornene

During the 1970s workers at the NASA Lewis Research Center developed a convenient method for the preparation of polyimides. It consisted of the use of a reactive norbornene end group attached to an imide oligomer. The technique avoided the construction of the imide group during MW advancement and was related to early work of Lubowitz [37].

One of the earliest systems was known as P13N with the "P" designation referring to polyimide, 13 representing a Mn of approximately 1300 g/mole and "N" for nadic end group. The reaction sequence entailed nadic anhydride (5 norbornene-2,3-dicarboxylic anhydride) reacting with methylene dianiline (MDA) and 3,3′,4,4′-benzophenone tetracarboxylic acid dianhydride (Eq. 4):

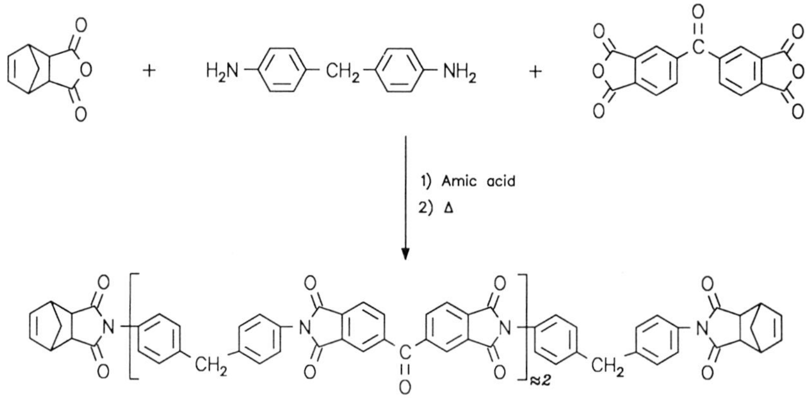

1) Amic acid
2) Δ

Equation 4

The resulting amic acid intermediate is cyclodehydrated to a soluble prepolymer whose MW is controlled by the relative ratio of reactants. The convenience of this method is that a relatively low MW polyimide (oligomer) is obtained at a reasonably low temperature to allow the release of volatiles.

The P13N species exhibited good processing when compared to the CPI method but the resulting cured resin was porous and brittle. Curing occurs at elevated temperatures (275 °C) by a combined addition reaction of the maleimide end group and the cyclopentadienyl group that are generated by a reverse Diels-Alder reaction (Eq. 5):

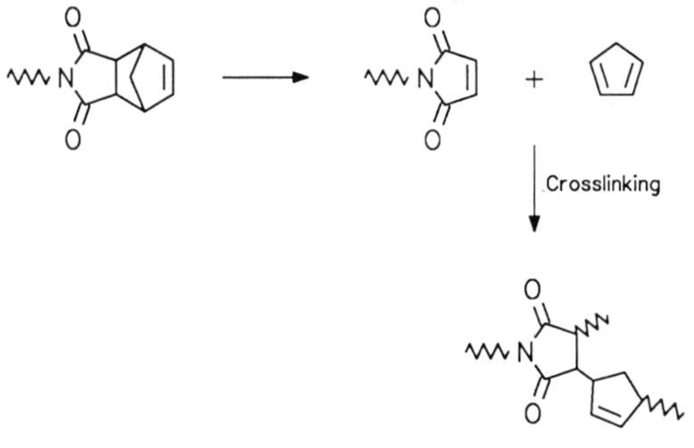

Crosslinking

Reverse Diels–Alder systems

Equation 5

A dense network of relatively short, stiff chains results. Serafini [38] at NASA Lewis extended the norbornene concept by developing a highly successful second generation API product that was analogous to P13N. This product became known as PMR-15 or "Polymerization from Monomeric Reactants" with 15 representing an average Mn of 1500 g/mole. By reacting the dimethyl ester of 3,3′,4,4′-benzophenone tetracarboxylic acid with MDA and the monomethyl ester of 5 norbornene-2,3-dicarboxylic acid, a soluble polyamic prepolymer composition (in methanol) is obtained with excellent tack and drape for impregnating fibrous reinforcing agents (Eq. 6):

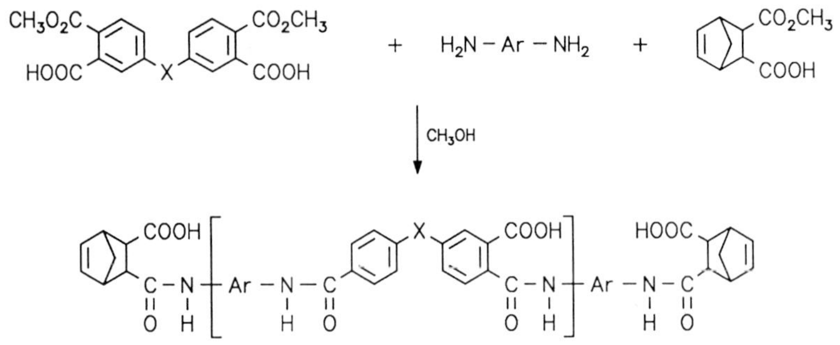

API			
PMR 15	X = CO	Ar = p–$C_6H_4CH_2C_6H_4$–	M_n = 1500
PMR–II–30	X = $C(CF_3)_2$	Ar = p–C_6H_4–	M_n = 3000
LARC 160	X = CO	Ar = Jeffamine	M_n = 1600

Equation 6

Features of PMR included ease of devolatilization, and slightly improved toughness attributable to a more flexible crosslinked network. A multi-step heating cycle is necessary to initially remove methanol/water and facilitate an in-situ cyclodehydration to the norbornenyl low MW imide oligomer. This cyclodehydration is referred to as "B staging" and is reminiscent of the more commonly B staging of phenolics. A large data base has been generated for PMR-15 type composites comprising different cure cycles and different types of reinforcing agents [39–42].

Although PMR-15 is one of the preferred high temperature resin systems for high performance composites, it is not without problems. Some contaminates such as the trimethyl ester (XV), amide (XVI) and/or the imide (XVII) are formed during storage of methanolic solutions of PMR-15.

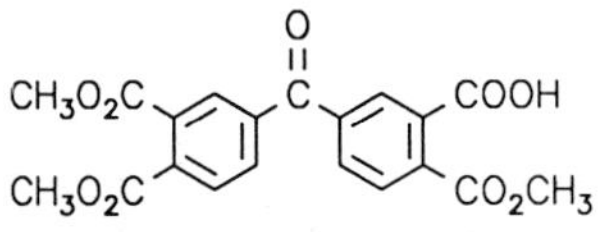

XV

XVI

XVII

Roberts and Vannucci [43] established the presence of the impurities by HPLC. More detailed separation and identity of most components for improved quality control monitoring has been accomplished by Escott [44] (Fig. 2). The identity of many of these components has facilitated an understanding of solution and prepreg variability due to the formation of undesirable intermediates during storage.

Besides critical quality control monitoring and prepreg variability of PMR-15, microcracking during thermal cycling is reported to occur. Continued thermal cycling (2000 cycles in air from − 18 to 232 °C) results [45] in thermally induced microcracks. Best microcracking resistance [45] occurs with reduced cure temperatures and lower fiber volumes with random carbon mat surface plies.

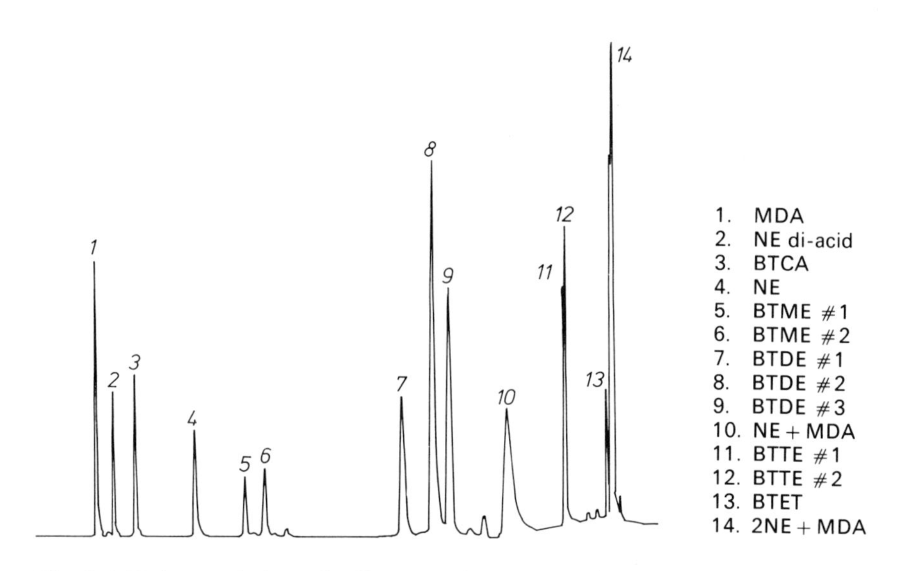

Fig. 2. MDA = methylene dianiline, BTCA = tetra carboxylic acid of BTDA, NE = nadic ester, BTME = monomethyl ester of BTDA, BTDE = dimethyl ester of BTDA, NE/MDA = 1:1 reaction product of NE and MDA, BTTE = trimethylester of BTDA, BTET = tetramethylester of BTDA

Wilson, [46] in a recent article summarizing PMR-15 processing, identified a number of process limitations such as quality control reliable methods, variability of batch to batch prepreg, microcracking caused by thermal cycling, worker and environmental hazards, and high temperature processing.

Despite these problems, PMR-15 composites particularly with carbon fiber are widely used in many high temperature applications such as F404 engine duct, ultra high speed fan blades, inner cowl, swirl frame, other engine and nacelle components as well as missiles and the shuttle orbiter.

PMR-15 is (Table 4) commercially available as a low viscosity 70% methanolic solution of monomers or as a powder known as Hycomp M-100 [47]. Prepregs of PMR-15 are available from SP Systems, Hexcel, NARMCO (BASF) and U.S. Polymeric.

PMR-15 has also been evaluated as a carbon fiber sizing agent [48] for T40R fibers. Data comparison of retention of interlaminar shear and isothermal ageing of several CF sizes indicated the PMR-15 sized composition (all PMR-15 resin matrix) required 5400 hours at 288 °C for a 5% weight loss. A corresponding partially fluorinated polyimide (with anticipated good thermal stability) size on T40R fibers survived only 4000 hours for 5% weight loss.

The success of PMR-15 prompted second generation type PMR materials with improved thermooxidative stability over PMR-15 [49]. Systems based on nadic ester, phenylene diamine mixture and the dimethyl ester of 4,4′ hexafluoroisopropylidene-bisphthalic acid with MW varying from 1300 to 5000 have been developed. These materials are PMR-II-13, PMR-II-30 and PMR-II-50 depending on MW (Eq. 6).

Solutions of monomers as 50% in methanol, fully imidized powders or prepregs are available from Dexter Hysol Composites.

Table 6 provides a weight loss comparison of these second generation materials with PMR-15. With increasing MW, PMR-II-50 exhibited the lowest weight loss at 371 °C after 300 hours as compared with PMR-15.

The brittleness of PMR-15 can be minimized by introducing a thermoplastic composition or a ductile phase into the brittle PMR-15. The combination of PMR-15 and NR 150 B2 (see thermoplastic polyimide section) yields a tougher, more microcrack resistant, high temperature resin known as LARC RP40 [50].

Table 6. Weight loss of molded neat PMR type resin

Resin	T_g (°C)[a]	T_g (°C)[b]	Percent weight loss after:	
			300 h/1 bar	75 h/4 bar
PMR-15	370	388	18.0	18.2
PMR-II-13	368	381	13.0	12.3
PMR-II-30	345	368	8.0	6.4
PMR-II-50	340	355	5.5	5.0

[a] After 24 hour postcure at 371 °C
[b] After 50 hour exposure, 371 °C, 1 bar

The resulting composition, LARC RP40, is a semi-interpenetrating network (see IPN section) based on easily processable but brittle PMR-15 and a tough but difficult to process thermoplastic polyimide, NR 150-B2. The fracture toughness of LARC-RP40 is increased fourfold over PMR-15. Thus the resulting semi-IPN LARC RP40 is tougher, higher in T_g and resistant to microcracking.

Variations of PMR-15

Extension of the P13N and PMR-15 technology by NASA Langley researchers resulted in relatively solvent free systems, LARC-160 and LARC-13. The former was a melt processable oligomer from the reaction of diethyl ester of 3,3′,4,4′ benzophenone tetracarboxylic acid, nadic acid and a liquid eutectic amine mixture of methylene dianiline, known as Jeffamine (Eq. 6 Ar = Jeffamine). With a Mn of about 1600 g/mole, LARC 160 is a melt processable composition which processes like an epoxy resin but the thermooxidative stability of LARC-160 is lower than PMR-15 [51].

The LARC-13 system is based on the reaction of 3,3′,4,4′-benzophenone tetracarboxylic dianhydride, 3,3′ MDA with nadic anhydride terminating the amic acid [52]. The stoichiometry is maintained at an Mn of 1300 g/mole. Modification of this methodology has been extended to thermoplastic polyimides known as LARC TPI (see Thermoplastic Resins).

Although the ease of processing PMR type oligomers is facilitated by the nadic end group, it is also responsible for reduced thermooxidative stability of the PMR system due to the aliphatic nature of the nadic end group. A recent, [53] novel sequence reported by NASA Lewis workers that avoids the nadic end group is the use of the [2.2] paracyclophane end group. The purpose of the cyclophane method is to reduce the aliphatic content of the reactive end group of the PMR type (API) and yet process in a manner analogous to structures with a nadic end group at elevated temperatures. The cyclophane ring thermally cleaves into a diradical species [54] that spontaneously polymerizes into poly-*p*-xylylene (Eq. 7):

$$\xrightarrow{\Delta} \left[\cdot H_2C-C_6H_4-CH_2\cdot \longleftrightarrow H_2C=C_6H_4=CH_2 \right] \longrightarrow \left[CH_2-C_6H_4-CH_2 \right]_n$$

Equation 7

Either an amino or amino-carboxamide group was introduced into one of the phenyl rings of the paracyclophane unit. In a manner analogous to the nadic end group of PMR-15, the paracyclophane unit was the terminus of the oligomeric API. The reaction sequence is as follows (Eq. 8):

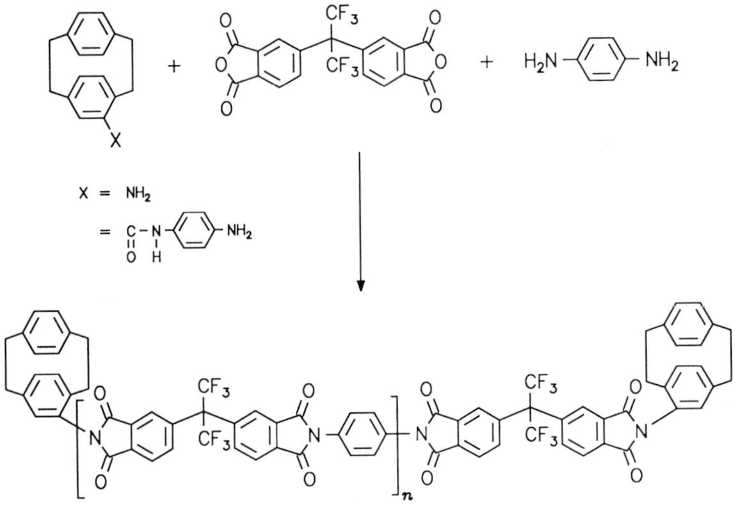

X = NH_2

= $C(=O)-NH-C_6H_4-NH_2$

XVIII

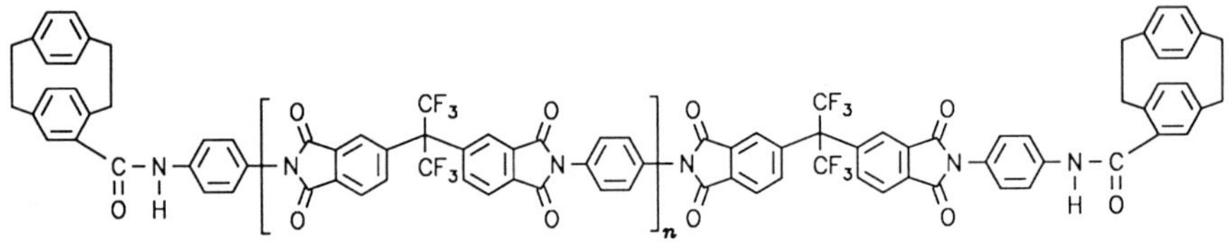

XIX

Equation 8

Analyses of either oligomer (XVIII and XIX) by DSC indicated that polymerization occurred between 300–350 °C. Thermal gravimetric analyses (TGA) of the materials in air indicated an onset of decomposition of about 550 °C for either material. Improved thermooxidative stability is observed for the paracyclophane API system as compared to PMR-15.

2.2.4.2 Acetylene

The highly successful technique of appending a nadic end group to polyimide oligomers spawned other reactive-end groups oligomer research. In the early 1970s acetylene terminated imide research led to the development of HR600 [55, 56]. It was later commercialized as the Thermid MC600 family of products [57]. The preparation of Thermid MC600 is as follows (Eq. 9):

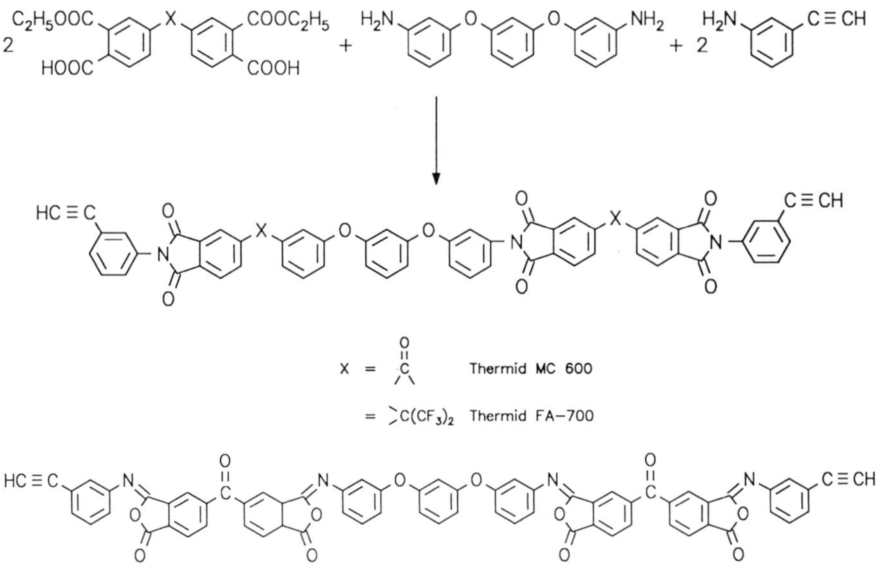

X = C=O Thermid MC 600

= >C(CF$_3$)$_2$ Thermid FA–700

Thermid IP 600

Equation 9

A lengthy, high temperature curing cycle is required for transforming thermid MC600 into a fully crosslinked material. Properties of Thermid MC600 are contained in Table 4. An acetylene terminated isoimide (isomeric imide), known as Thermid IP600, is also commercially available from National Starch [57] and is reported to be more soluble and more easily processable than Thermid MC600, but with similar mechanical properties on curing.

More recently Thermid FA-700 [X = $C(CF_3)_2$] has been reported by Capo and Schoenberg [58]. The newer composition with a hexafluoro isopropylidene linkage instead of a carbonyl group that links the phthalimide units exhibits better thermal and oxidative stability as compared to either Thermid MC600 or IP600.

Other acetylene terminated oligomers, besides polyimide, have been reported and include, aryl sulfone oligomers [59], aryl ether oligomers [60] and phenylene oligomers [61]. To date, none of these materials has been commercialized.

A closely related API with an allyl group attached to the nadic end group has been developed by Renner and coworkers [62, 63] (Eq. 10).

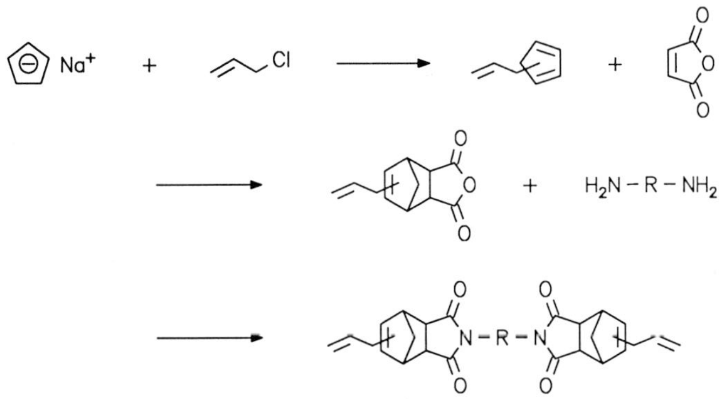

R = p–$C_6H_4CH_2C_6H_4$ XX

R = –$(CH_2)_6$– XXI

Equation 10

The “allyl nadic imides” are more reactive than the unsubstituted nadic oligomers, are lower melting oligomers, and soluble in common organic solvents [64]. Compounds XX and XXI as well as XXII on curing exhibit T_gs in excess of 300 °C. These materials co-react with bismaleimides (see next section) and lead to compositions with reduced brittleness without any loss in mechanical or thermooxidative properties.

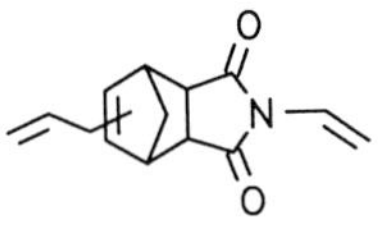

XXII

2.2.4.3 Cyclobutene

A rather unusual reactive end group sequence is the use of benzocyclobutene terminated aryl imide oligomers [65, 66, 67] (Eq. 11). Arnold and coworkers have described some novel systems with T_gs (cure) ranging from 240 to 280 °C (Table 7). Cure of the benzocyclobutenes is proposed to occur via the thermal electrocyclic ring opening of cyclobutene [68] (Eq. 12).

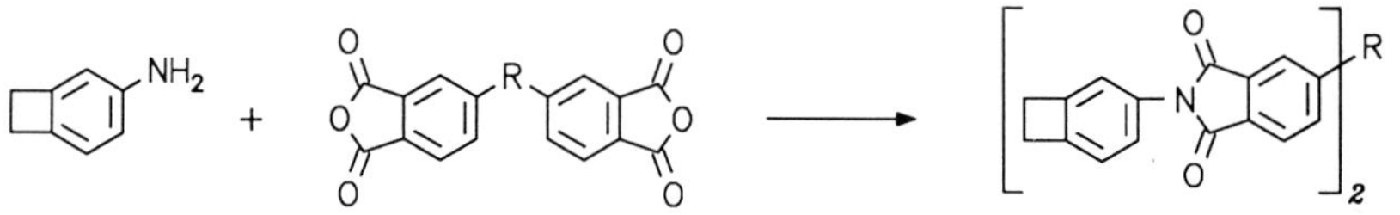

Equation 11

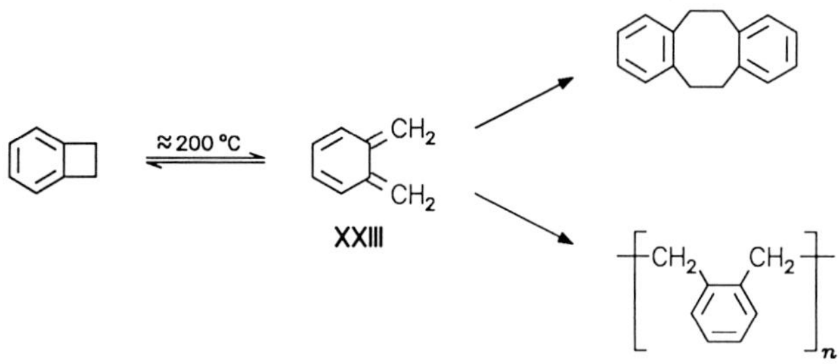

≈200 °C

XXIII

Equation 12

Table 7. Glass transition temperatures, T_g, of benzocyclobuteneimides (DSC)

R	T_g, °C
$>C(CF_3)_2$	287
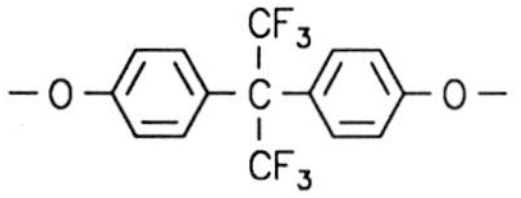 $-O-C_6H_4-C(CF_3)_2-C_6H_4-O-$	247 251 (DMA)
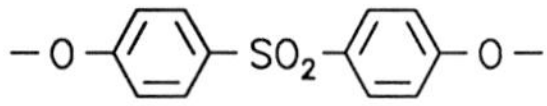 $-O-C_6H_4-SO_2-C_6H_4-O-$	240

The highly reactive *o*-xylylene (XXIII) (analogous to thermal cleavage of paracyclophane to *p*-xylylene) can undergo cycloaddition forming an eight membered ring (dimerization) or linear polymerization. In the presence of suitable dienophiles, such as acetylene or maleimide, benzocyclobutene will add to these reactive substituents via a Diels-Alder reaction (Eq. 13):

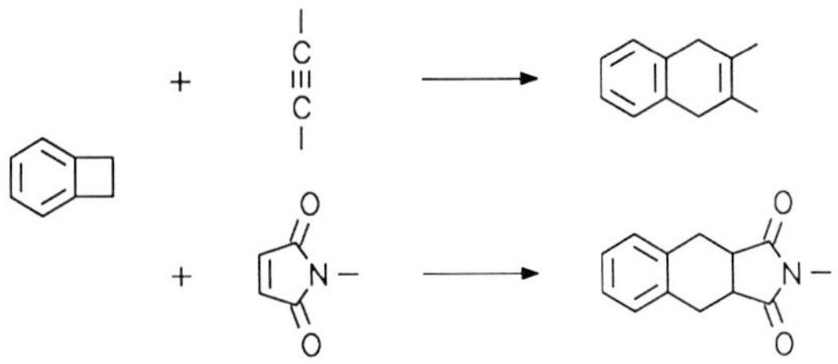

Equation 13

Unusually high T_g benzocyclobutene imide and benzoxazole oligomers have been prepared as precursors for molecular composites (see later in Molecular

Composites section). Oligomers 6F-BCB and PBO-BCB on curing exhibited T_gs of greater than 400 °C by Torsion Pendulum [Impregnated Cloth] Analysis [69, 70] (Eq. 14).

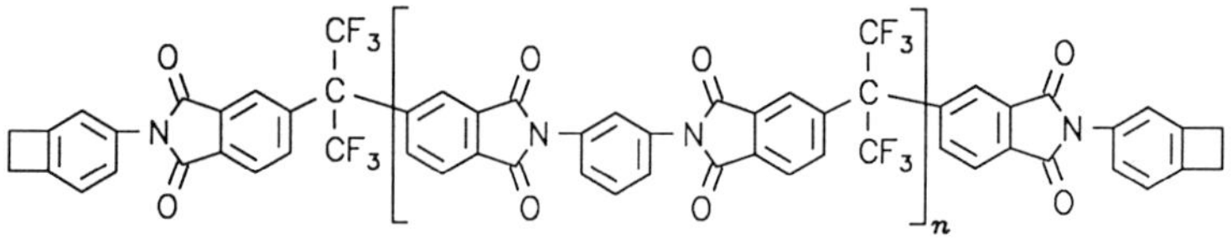

6F–BCB

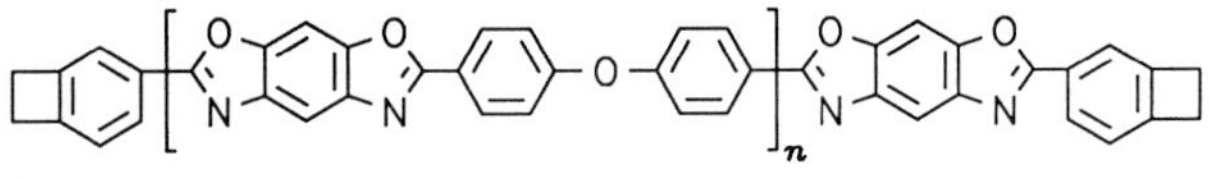

PBO–BCB

Equation 14

Arnold [71] has combined benzocyclobutene imide with reactive dienophiles within the same oligomer. Various synthetic strategies have resulted in several benzocyclobutene-dienophile compounds. The versatile benzocyclobutene group can accommodate facile coreaction with the reactive dienophile such as the substituted acetylene or maleimide group as well as undergo polymerization.

The benzocyclobutene when coreacted with bismaleimides (BMI) provides significant improvement in the thermooxidative stability of BMIs. When benzocyclobutene [Table 7, R = $C(CF_3)_2$] is combined with BMI of MDA [72], only 15% weight loss occurs after 200 hours at 343 °C in air. In the absence of the benzocyclobutene, BMI MDA retained only 3% of original weight under similar conditions.

2.2.4.4 Bismaleimides

A closely related reactive compound contained within API is the bismaleimide family of materials [35, 73]. The bismaleimide (BMI) resin system can vary from a crystalline difunctional compound to an oligomeric amorphous material with difunctionality. In recent years many BMI resins and modifications have been developed primarily due to good drape and tack and "epoxy-like" cure conditions resulting in resins with higher temperature characteristics. BMIs are quite versatile, capable of thermal polymerization as well as copolymerization under nucleophilic conditions in an alternating manner with amines. BMIs possess thermal capabilities between 175–230 °C (epoxies 120–175 °C). They are readily prepared by reacting aromatic diamines with maleic anhydride (Eq. 15):

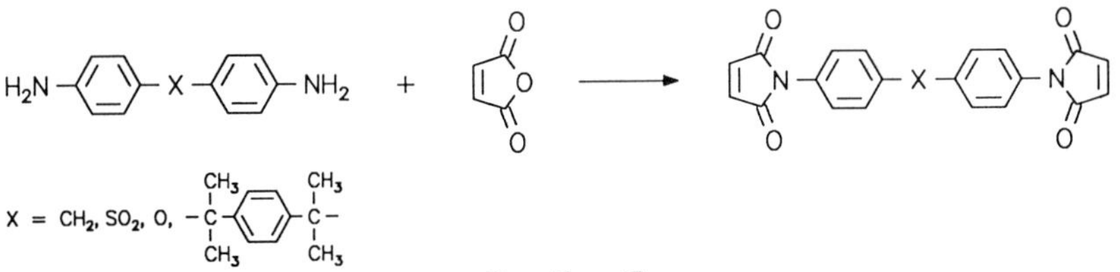

$$X = CH_2, SO_2, O, -C(CH_3)_2-C_6H_4-C(CH_3)_2-$$

Equation 15

Hence a wide variety of BMIs are possible depending on the aryl diamine precursor. The most popular BMI is based on 4,4′ MDA ($X = CH_2$) and is a major component in many commercial products supplied by Rhone Poulenc, Ciba Geigy, and others.

Mechanical properties of BMI resins are contained in Table 4. BMI resins are a very rigid system with a corresponding high degree of crosslinking, hence brittleness. The latter characteristic leads to microcracking and high shrinkage.

Early modifications of BMI resins involved multicomponent systems consisting of BMI and diamine extended BMI oligomers (Eq. 16):

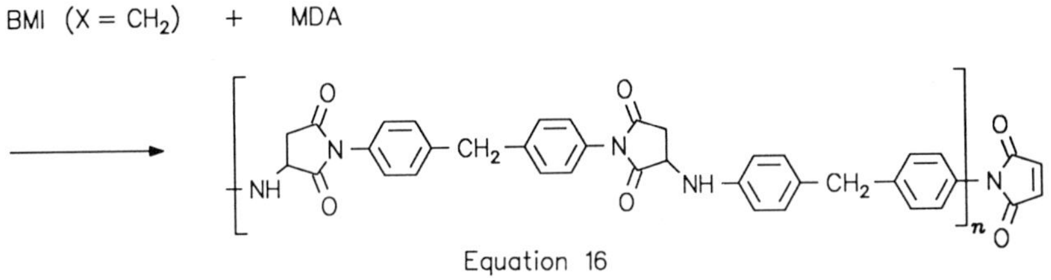

BMI ($X = CH_2$) + MDA

Equation 16

Many amine chain extended resins are hot melt processable for prepregging onto reinforcing fibers. Some resins have a long pot life of 10 hours at 100 °C as well as lengthy cure cycles. Applications include aircraft engine motor cases, nose cones for jet engine, and other high temperature application areas.

Polyfunctional Maleimides

Polyfunctional maleimides can be prepared via a multi-step sequence beginning with terephthaldehyde [74] (Eq. 17).

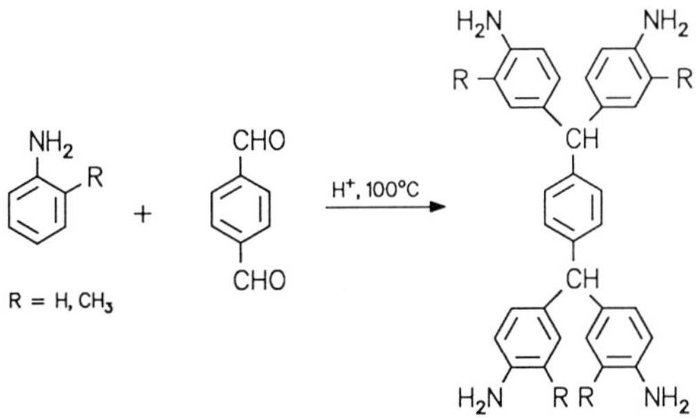

R = H, CH_3

H^+, 100°C

Equation 17 continued

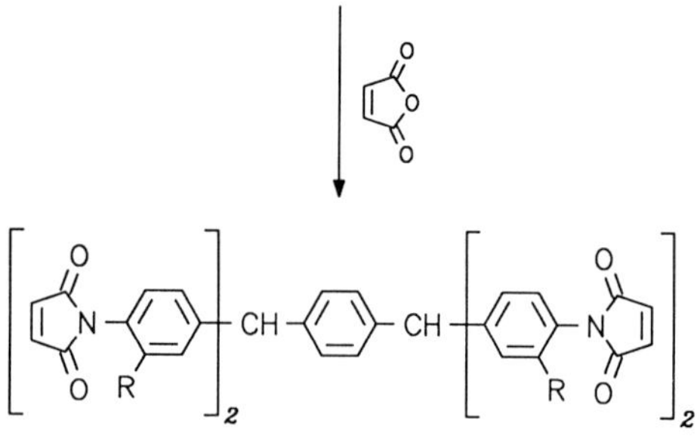

Equation 17

If either pure aniline or *o*-toluidine is reacted with terephthaldehyde, high temperature softening (200 °C) polyamines are obtained with limited solubility in organic solvents. By using a mixture of aniline/*p*-toluidine, low softening point (110–130 °C) soluble polyamines can be transformed into the corresponding polyfunctional maleimides. Polyfunctional maleimide cured with 3,3′ dimethyl 4,4′ diamino diphenyl methane exhibited a T_g of 420 °C.

Oligomeric BMI

Diamine terminated oligomers can be transformed into oligomeric BMIs and then crosslinked into network polymers. Kwiatkowski and coworkers [75] observed a rapid decrease in both T_g and shear modulus as the oligomer unit of polysulfone was increased from 1 to 2 during maleimide transformation and curing (Eq. 18):

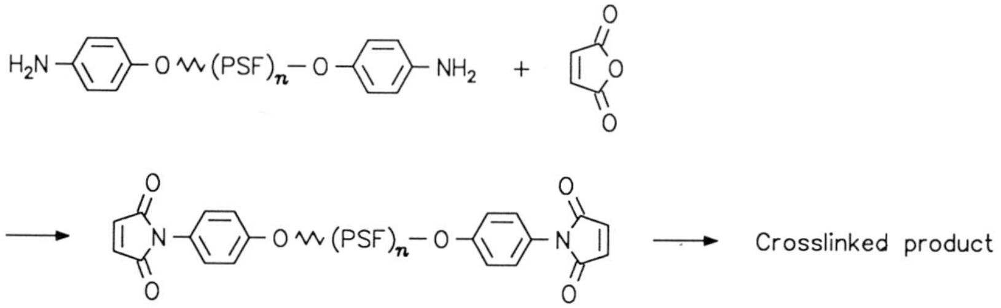

$$H_2N-C_6H_4-O\sim(PSF)_n-O-C_6H_4-NH_2 \; + \; \text{maleic anhydride}$$

$$\longrightarrow \text{BMI}-C_6H_4-O\sim(PSF)_n-O-C_6H_4-\text{BMI} \longrightarrow \text{Crosslinked product}$$

n	1	2
T_g, °C	239	185

Equation 18

As the oligomer MW was increased, the BMI end capped material on crosslinking resembled polysulfone (T_g PSF = 190 °C) in overall properties. In extending this technique McGrath and coworkers [76] examined different methods of appending the maleimide group to the terminus of polyaryl ether ketone oligomers and determined the most convenient method consisted of coreacting polyaryl ether ketone monomers with maleimide terminated phenols (Eq. 19):

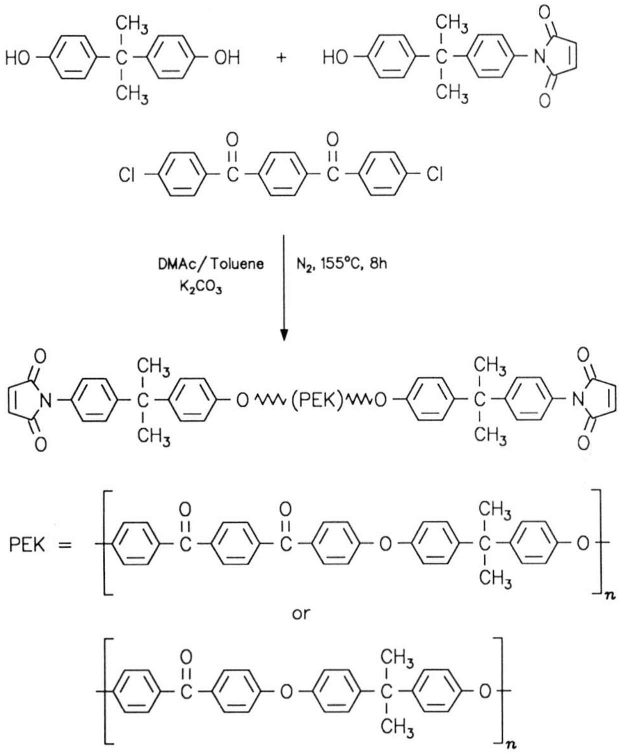

DMAc/Toluene K_2CO_3 | N_2, 155°C, 8h

PEK =

or

Equation 19

The already assembled maleimide phenol avoids the question of whether residual uncyclized amic acid can be present when amino terminated oligomers are reacted with maleic anhydride. These oligomeric maleimide terminated polyaryl ethers on crosslinking yielded solvent resistant materials and exhibited improved fracture toughness.

Toughened BMIs

One of the serious shortcomings of the BMI cured system was that the high crosslink density facilitated microcracking. These BMI systems were brittle, and extensive programs were undertaken to reduce BMI brittleness by impact modification or through the use of ductile resins.

Various methods have been examined as a means of toughening or improving the impact of BMI resins. Varma [77], Shaw and Kinlock [78] introduced amine and carboxyl terminated rubbers respectively and observed increased short beam shear strength (Varma) or fracture energy (Shaw) of the modified BMI systems. Recently Takeda [79] modified BMI resins with carboxyl terminated butadiene acrylonitrile rubbers. The morphology of the resulting resins by SEM and DMA indicated that the rubbers improved toughness of the BMI resins without significant reduction in heat resistance.

Besides rubber modification, the use of addition copolymerization of BMI with styrene, hydroxyethyl methacrylate has yielded tougher materials with fracture toughness values of 170 J/m^2 as compared to 30–50 J/m^2 for unmodified BMI [80].

In the same study the authors demonstrated that the fracture energy was increased between 300 and 1000 J/m^2 when the free radical copolymerization was carried out with a reactive group terminated nitrile rubber (amine, carboxylic acid or vinyl).

The use of co-reactive rubbers [81] or comonomers unfortunately results in a corresponding reduction in T_g, compressive strength as well as modulus of the modified BMI.

A more facile method for improved BMI toughness has been co-reaction of alkenyl or allyl substituted phenyl ether oligomers with BMI.

In a series of papers Stenzenberger [82] showed that a significant increase in BMI toughness (high fracture energy) can be obtained by coreaction of BMI with oligomeric allyl phenyl compounds or alkenyl terminated polyaryl ether ketones or sulfones (Eq. 20).

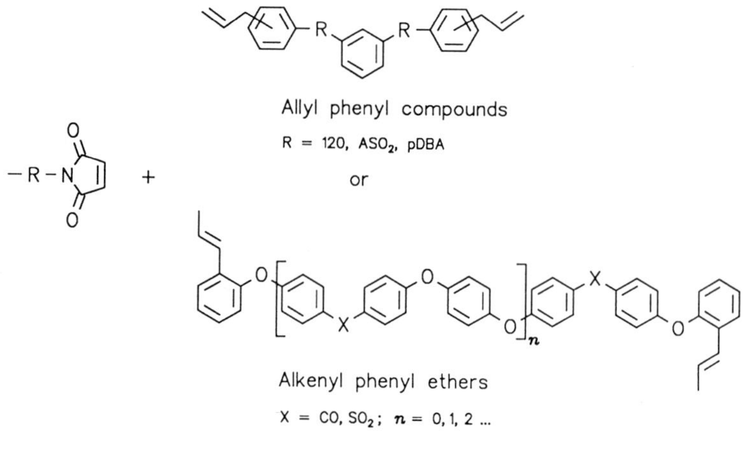

Allyl phenyl compounds

R = 120, ASO_2, pDBA

or

Alkenyl phenyl ethers

X = CO, SO_2; n = 0, 1, 2 ...

Equation 20

With either coreactive species, Stenzenberger obtained modified BMI materials with G_{IC} values of 400–500 J/m^2 versus a value of about 50 J/m^2 for the neat BMI resin. Besides toughness, the modified BMIs exhibited improved mechanical properties but a corresponding reduction in T_g of 250–260 °C (vs > 300 °C).

Later Stenzenberger [83] combined the modified BMI (BMI + alkenyl ether ketone) with a thermoplastic component as a means of introducing a ductile phase into the moderately rigid modified BMI by using the semi-IPN method. The semi-IPN technique has been utilized in thermoset systems such as epoxies [84], cyanates [85] and neat BMI resin [86].

According to Stenzenberzer, a co-continuous network of phase separated components (0.05–0.5 μm domain size) of high performance thermoplastic resins such as polysulfone, polyetherimide, or polyhydantoin was essential for achieving significant toughness without a corresponding compromise in high temperature characteristics of the BMI alkenyl ether ketone systems. Fracture toughness (G_{IC}) was significantly increased by 4 to 5 fold through the introduction of 20–30% of high performance thermoplastic resin. These same improved properties carried over to the carbon fiber composites.

Ternary compositions of BMI alkenyl ether ketone/PEI exhibited phase separation while no phase separation could be observed for BMI/alkenyl ether ketone/polyhydantoin. The latter composition possessed excellent fiber adhesion with a large fracture surface area typical for the presence of a ductile resin, (Table 4).

Thus ternary BMI systems as semi-IPNs offer promise as high temperature, moderately tough, crosslinkable materials that are somewhat more easily processable due to the low MW reactive oligomers (BMI/alkenyl ether ketone) that aid in reducing the high melt viscosity of the thermoplastic resin.

2.2.5 Cyanate Esters

Another thermally reactive system which is attracting considerable interest is the cyanate ester substituent contained within a compound or an oligomeric material. Cyanate esters resemble epoxy or BMI substituents because they undergo crosslinking or a curing reaction without the release of volatiles. Cyanate esters possess higher performance characteristics when compared to epoxies due to the former's better heat resistance and strength. Cyanate esters are more easily processed than BMIs. They can be used in filament winding, compression molding and the customary prepreg processing.

Recently cyanate esters of Bisphenol A (XXIV) and related oligomers (XXV) were commercialized [87] (Table 8). Oligomerization occurs through trimerization of the cyanate group. Either monomer or oligomer undergoes polymerization or crosslinking with selected catalysts such as metal carboxylates or chelates of Cu, Zn, or other transition metals in conjunction with an active hydrogen (nonyl phenol, imidazoles) co-catalyst (Table 4).

Table 8. Commercial cyanate esters [87]

	XXIV	XXV	
Physical state	Crystals	Semi-Solid	Solid
M.P. (°C)	79	–	–
Melt visc (cps.)	–	750 (82 °C)	3000 (149 °C)
%Trimer	–	30	50
Cyanate equivalent	139	202[a]	270[a]

[a] Approximate by IR method

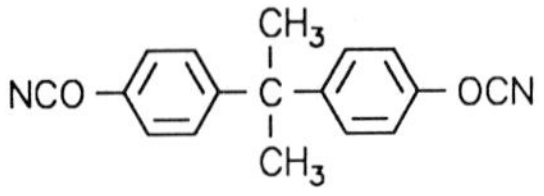

XXIV

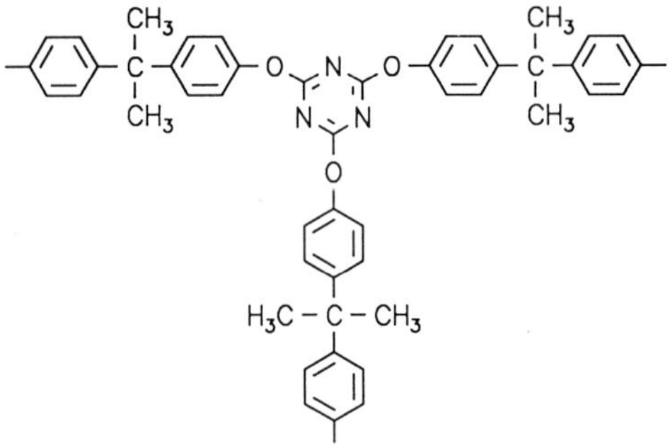

XXV

Newer polycyclic aryl cyanate esters based on a multi-step synthesis have been reported [88] (Eq. 21):

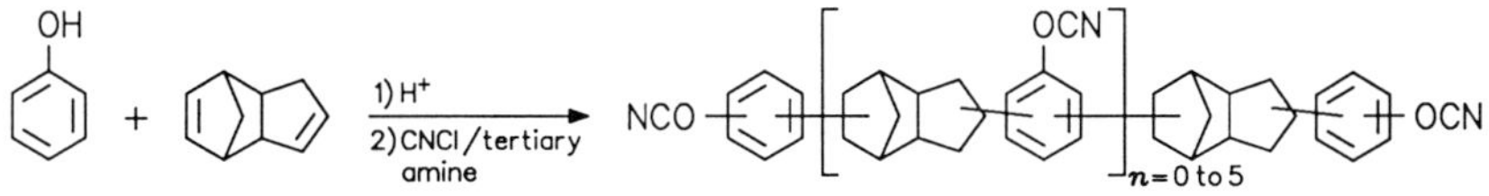

1) H^+
2) CNCl/tertiary amine

$n = 0$ to 5

XXVI

Equation 21

Comparison of crosslinked XXVI versus XXIV and another dicyanate XXVII indicated that XXVI was superior to XXIV and XXVII in lengthy exposure to boiling water, dilute HCl and 40% NaOH.

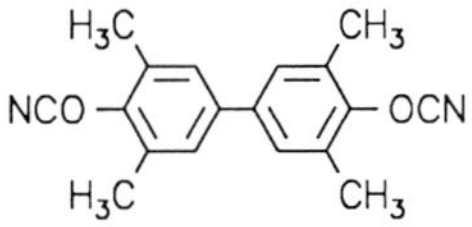

XXVII

Properties of the rigid crosslinked system XXVI are slightly different from DDS + multifunctional epoxies and BMI materials. Yet XXVI is less water sensitive at elevated temperatures. The rigid, hydrophobic structure combined with the high crosslink density results in thermally stable, moisture resistant compositions with good retention of hot/wet mechanical properties (Table 4).

The appending of a cyanate group to hydroxyl containing materials has been extended to phenolic resins by Prevorsek and coworkers [89, 90]. The reaction of cyanogen halide with phenolic resin leads to an intermediate polyfunctional cyanate containing resin which thermally trimerizes/crosslinks into a phenolic triazine (PT resin) matrix system (Eq. 22). A more stable polymer with improved thermal and oxidative stability (as compared to phenolic resin) is obtained with a T_g greater than 300 °C, good retention of mechanical properties above 300 °C and a char yield of 70% at 1000 °C. The authors claim that the PT resin is comparable to PMR-15 and other related high performance thermoset systems for high temperature use above 316 °C.

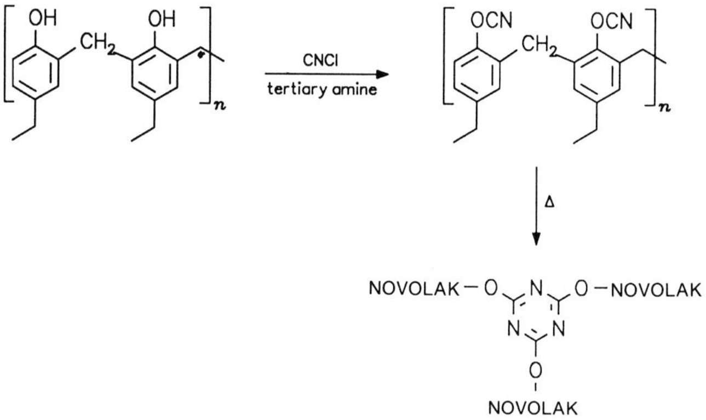

OH CH2 OH n — CNCl, tertiary amine → OCN CH2 OCN n — Δ → NOVOLAK—O—(triazine)—O—NOVOLAK, O—NOVOLAK

Equation 22

Co-reaction of cyanate esters with epoxies, BMI resins and high performance thermoplastic resins has expanded the utility of cyanate esters. The bismaleimide triazine (BT) resins made by co-reaction with BMI have been available since 1976 and are widely used in electrical laminates [91]. Recent versions of the BT resins incorporate brominated epoxy resins as a means of providing flame retardancy [92]. DSC and FTIR studies have shown that the onset of cyclotrimerization of cyanate ester to a triazine structure occurs at 150 °C. At a higher temperature epoxy homopolymerization and BMI polymerization occur as well as a co-reaction of epoxy with cyanate to form an oxazoline ring [93] (Eq. 23):

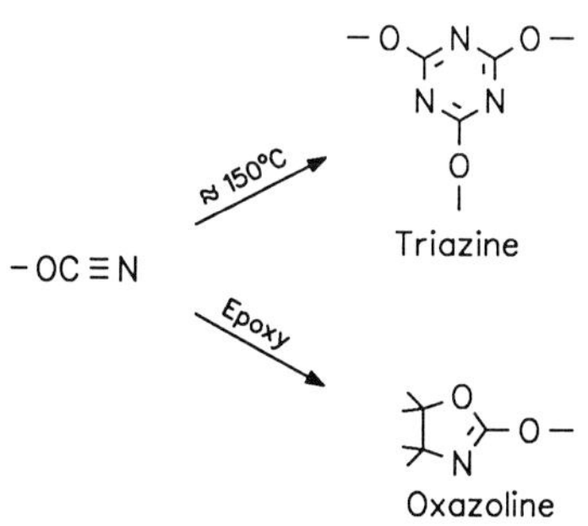

-OC≡N — ≈ 150°C → Triazine; — Epoxy → Oxazoline

Equation 23

Semi-IPNs are obtained by thermally curing cyanate esters in the presence of high performance thermoplastic resins. Thermoplastic resins such as polycarbonate, polysulfone, polyether imide, and polyarylate were co-reacted with Bisphenol A dicyanate esters XXIV [94, 95]. Greater strength and toughness for the thermoset portion is attributable to the high performance thermoplastic resin while ease of processing and high heat resistance is provided to the thermoplastic portion by the cyanate ester.

The presence of both cyanate and maleimide functionalities contained within the same molecule (XXVIII, XXIX) has been reported by Hefner [96].

Besides co-curing with XXIV or BMI/MDA, XXVIII and XXIX are copolymerized with selected monomers such as styrene or acrylates via radical conditions. The hybrid cyanate/maleimide compounds XXVIII and XXIX are quite versatile undergoing either free radical copolymerization or co-reaction with dicyanate esters or BMI resins.

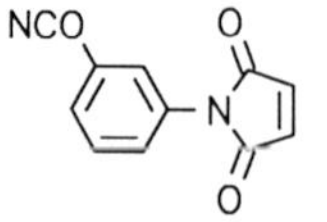

XXVIII 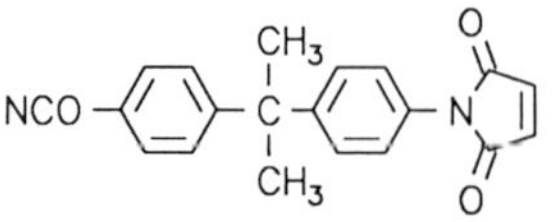 XXIX

2.3 Summary of Thermoset Resin Performance and Future Prospects

The low viscosity and low MW oligomeric TS resin compositions continue to dominate in the design of final composite components. Many of these TS systems benefit from the large data bases that have been developed since the inception of reinforced composites in the late 1930s. With the emergence of high strength, high temperature TS resins with significantly improved fracture energies (improved damage tolerance) due to the use of multi-phase or IPN, TS systems will continue to be the preferred resin system because of the dedicated TS fabrication equipment that most composite part manufacturers presently possess.

2.4 Thermoplastic Resins

The introductory comparison of matrix resins described the advantages and the disadvantages of thermoset and thermoplastic resins as matrix resins for composites. In discussing thermoplastic (TP) resins it is also necessary to contrast the strengths and limitations of TP as they are utilized in composites as matrix resins. It is generally recognized that the use of TP resins will result in low cost manufacturing of composites. Factors which contribute to low cost manufacture are indefinite prepreg stability without refrigeration, a relatively fast processing cycle, easy quality control/assurance, and the potential to reprocess the molded component to correct imperfections. Moreover favorable TP characteristics such as toughness with accompanying improved damage tolerance provide additional incentive for considering TP resins. Thermoplastics resins, however, are either amorphous or crystalline and the resulting polymer morphology relates to solvent sensitivity or resistance. Amorphous resins are solvent sensitive whereas crystalline resins are highly solvent resistant. A uniform level of crystallinity must be attained during processing/fabrication of composites by controlling heating/cooling rates and indicating whether annealing conditions would be required. With a limited data base for TP resins, fatigue performance and creep behavior are additional TP resin characteristics which must be assessed. Nevertheless, the benefits of TP resins are sufficiently attractive that many military programs are being currently funded as well as "in-house" programs of TP manufacturers.

2.4.1 Polyaryl Ethers

In the mid 1960s, Farnham and Johnson [97] of Union Carbide showed that a unique thermally stable, oxidation resistant and rigid thermoplastic polyaryl ether sulfone resin system could be obtained by nucleophilic displacement of 4,4′ dichloro diphenyl sulfone by disodium bisphenate. Flexibility or toughness was provided by both the ether linkage and the isopropylidene group. These moieties improve melt characteristics whereby the polymer is more easily processed at lower temperatures. During this same period, ICI and 3M investigators reported the use of an electrophilic substitution process (Friedel-Crafts conditions) for the preparation of polyaryl sulfones. Of the two competing substitution processes, only the nucleophilic displacement process was commercialized to manufacture polyaryl sulfones.

Polyaryl ethers (PAE) can be prepared by a nucleophilic displacement of activated diaryl halides by alkali metal phenoxides. A vast array of PAE are available by this technique [98–102]. An alternate technique which is the Friedel Crafts method can be utilized to prepare polyaryl ether ketones and polyaryl sulfones.

2.4.1.1 Nucleophilic Displacement Polymerization

(1) Activated Halides

The facile technique of preparing PAE involves the use of diaryl halides which are activated by groups such as sulfone, carbonyl or nitrile. Displacement of these dihalo compounds is facilitated by alkali metal phenoxides in an aprotic polar solvent. The ease of halide leaving groups is $F \gg Cl > Br$. Basicity or nucleophilicity of the phenoxide is also important.

Table 9. Mechanical properties of thermoplastic resins

Resin	T_g (°C)	T_m (°C)	Tensile strength (MPa)	Tensile modulus (GPa)	Fracture toughness (J/m^2)
Polylaryl ethers					
PSF (Udel)	190	—	76	2.2	3,200
PES (Radel)	220	—	83	2.4	5,500
PEK (Kadel)	140	340	95	3.6	—
PEEK	143	343	103	3.8	2,000
PEKK	156	338	102	4.5	1,000
Thermoplastic polyimides					
LARC-TPI MTC	255	—	136	3.7	2,770
Rogers	256	—	153	4.4	3,500
LARC-CPI	224	354	134	4.2	—
NR-150 B2	340		110	4.1	2,400
PEI	220		110	3.3	3,700
PI 2080	280		120	1.3	—
Torlon	275		193	4.8	3,400
Polyaryl sulfide					
PPS	85	285	65	3.8	210 (high crystallinity)
PPSS	218		92	—	—

The first reported use of this method was in 1965 and identified polysulfone as the first in a series of PAEs that was subsequently commercialized. The pioneering efforts of Johnson and Farnham of UCC led to the commercialization of polysulfone (Table 9) in 1967 (Eq. 24):

$$n\,Cl{-}C_6H_4{-}SO_2{-}C_6H_4{-}Cl + n\,NaO{-}C_6H_4{-}C(CH_3)_2{-}C_6H_4{-}ONa \xrightarrow[160°C]{DMSO} \left[{-}C_6H_4{-}SO_2{-}C_6H_4{-}O{-}C_6H_4{-}C(CH_3)_2{-}C_6H_4{-}O{-}\right]_n + 2n\,NaCl$$

PSF

Equation 24

Almost simultaneously Rose and coworkers [103] at ICI reported a somewhat similar nucleophilic displacement of 4,4′ dichloro benzophenone with disodio bisphenol A in DMSO with some CuO (Eq. 25):

DMSO, CuO

PEK

Equation 25

A low MW PAE was obtained because of the marginal reactivity of the dichloro benzophenone. Johnson [97] found that with 4,4′ difluoro benzophenone a high MW material could be obtained within 0.5 hours at 135 °C in DMSO. He developed the necessary new parameters required for the preparation of high MW PAE. Johnson further established that a dry, aprotic, polar solvent, DMSO or sulfolane, was necessary to dissolve the disodio bisphenate and maintain the "growing" polymer in solution in the presence of an inert atmosphere. The exact caustic concentration was necessary; an excess of caustic degraded the dihalide whereas reduced caustic resulted in lower polymer MW or incomplete reaction. A newer process improvement requiring the difluoroaryl can be conducted in toluene-NMP with anhydrous K_2CO_3 [104].

With the above conditions outlined, high MW amorphous PAE resins are obtained. Modified conditions are required when the resulting PAE is crystalline. Rose and coworkers developed the necessary solvent for highly crystalline PAE's, specifically those with carbonyl groups such as polyether ether ketone (PEEK) and polyether ketone (PEK). Diphenyl sulfone was identified as the preferred solvent. Diphenyl sulfone is chemically and thermally stable as compared to dimethyl sulfoxide [105]. Both monomers, 4,4′ difluorobenzophenone, 4,4′ dipotassio dihydroxy benzophenone, and diphenyl sulfone solvent are heated to 335 °C for 2–3 hours, cooled, and result in a hard solid which is pulverized, extracted with water/methanol to yield PEK in 98% yield (Table 9). The corresponding 4,4′ dichlorobenzophenone compound can be substituted for the difluoro compound but a significantly lengthy reaction cycle is necessary to obtain PEK with similar MW.

Similarly, the manufacture of polyether ether ketone (PEEK) consists of hydroquinone and 4,4′ difluorobenzophenone heated to 180 °C in diphenyl sulfone in a nitrogen atmosphere followed by the addition of a slight excess of powdered K_2CO_3 and a temperature increase to 200 °C for 1 hour (Eq. 26). Sequentially increasing the temperature to 320 °C yields a nearly white PEEK with I.V. from 0.79 to 0.92. The same procedure is applicable to polysulfones and polyketone/sulfone copolymers [106]. PEEK exhibits an excellent combination of properties such as thermal and hydrolytic stability, high strength and toughness, wear/abrasion resistance, and solvent resistance (Table 9). The investigation of thermoplastic composites particularly APC-2 which is Carbon

Fiber (AS-4) reinforced PEEK for aircraft structural applications [107] and other applications has led to the development of CF reinforced PEEK related resin systems closely related to PEEK and known as APC (HTA), APC (HTC), and APC (HTX) [108].

$$HO-C_6H_4-OH + F-C_6H_4-C(=O)-C_6H_4-F + K_2CO_3 \xrightarrow[150-300°C]{DPS} \left[-O-C_6H_4-O-C_6H_4-C(=O)-C_6H_4-\right]_n$$

PEEK

Equation 26

Ageing characteristics of APC-2 at 120 °C show no deterioration. However, at an elevated temperature of 220 °C and after 1000 hours in air the mechanical strength is reduced to 10% of the original value and is presumably attributable to surface oxidation.

APC-2 has been compared with PPS/CF system in Compressive Strength after Impact properties (CAI). (See Chapter 9, Damage Tolerant Composites). CAI values indicate that APC-2 is significantly better than PPS-CF or comparable to the toughest epoxy material [109].

Since the semi-crystalline PEEK has a T_g of 143 °C and T_m of 343 °C, other closely related polyether ketones/ether sulfone systems were developed by ICI and were either totally amorphous (HTA) with a high T_g of 260 °C or high temperature crystallizable material (HTC or HTX) [108, 110].

Properties of HTX suggest an improvement over APC-2 with HTX exhibiting a higher T_g of 210 °C and hence a proposed "service" temperature of 180 °C with excellent solvent resistance. Materials for the proposed USAF advanced tactical fighter (ATF) must retain structural integrity to 176 °C (350 °F) to meet many criteria of the aircraft. A description of the various parameters and methodology used to develop HTX is presented by Cogswell [111]. The procedure is actually a reversal in the usual route to composites which is customarily the synthesis of a polymer, followed by composite preparation and properties determination. The ICI group began with a composite and its end-use characteristics followed by interface requirements. These established the physical and fundamental characteristics of the polymer which then identified monomers and sequence requirements. As a result of the large data base related to PEK and PEEK, HTX structure emerged and justified this reversed procedure to resin structure.

McGrath [112] has prepared a modified PEEK with a ketimine functionality contained within the PEEK resin system (Eq. 27):

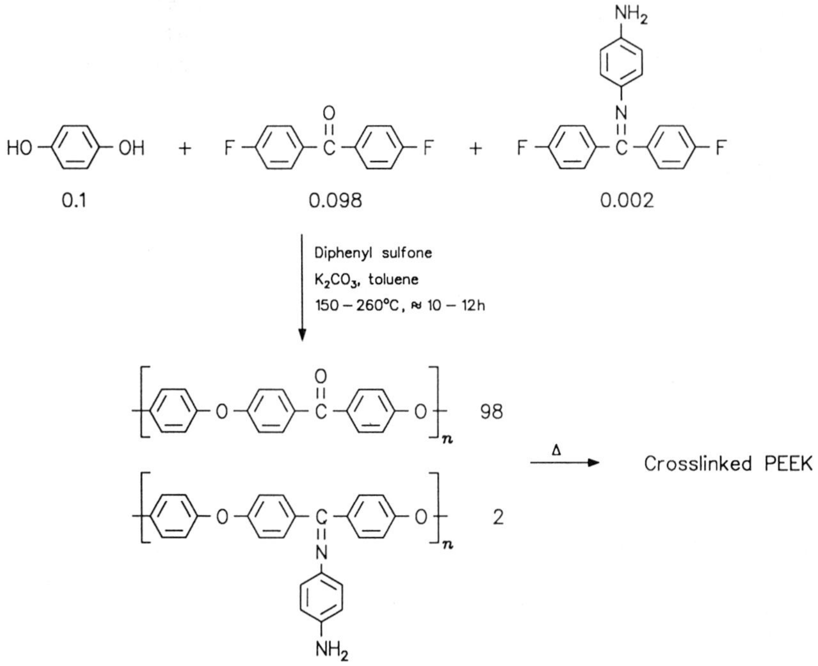

HO–C₆H₄–OH + F–C₆H₄–C(=O)–C₆H₄–F + F–C₆H₄–C(=N–C₆H₄–NH₂)–C₆H₄–F

0.1 | 0.098 | 0.002

Diphenyl sulfone
K_2CO_3, toluene
150 – 260°C, ≈ 10 – 12h

$[-C_6H_4-O-C_6H_4-C(=O)-C_6H_4-O-]_n$ 98

$[-C_6H_4-O-C_6H_4-C(=N-C_6H_4-NH_2)-C_6H_4-O-]_n$ 2

Δ → Crosslinked PEEK

Equation 27

The objective was to thermally crosslink the pendant amine group with the carbonyl group of PEEK. A slightly higher T_g was noted by DSC. It was also expected that crosslinked material would improve creep behavior of PEEK. TMA comparison of PEEK and crosslinked PEEK indicated that mechanical integrity of crosslinked material is maintained above its T_g due to crosslinked sites in the backbone.

(2) Activated Nitro Groups

Nucleophilic substitution of nitro groups activated by an imide function occurs as follows [113–115] (Eq. 28):

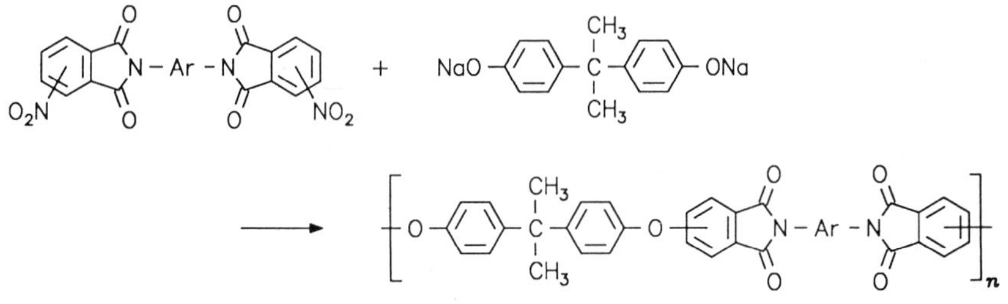

O_2N–(phthalimide)N – Ar – N(phthalimide)–NO_2 + NaO–C₆H₄–C(CH_3)₂–C₆H₄–ONa

→ $[-O-C_6H_4-C(CH_3)_2-C_6H_4-O-$(phthalimide)N – Ar – N(phthalimide)$-]_n$

Equation 28

The ether linkages of PEI provide ductility and processability while the imide linkage provides high heat resistance and strength. (See PEI in thermoplastic polyimide section).

2.4.1.2 Friedel–Crafts Polymerization

The Friedel–Crafts (FC) reaction consists of an electrophilic aromatic substitution of an acyl halide with an aromatic substrate. The earliest report of FC polymerization is by Bonner [116] and involved the co-reaction of aryl diacyl halides with diphenyl ether. Electrophilic acylation can lead to AB or AABB type polymers, and polymer characteristics depend on the solubility of the "growing" polymer in the reaction mixture.

A variety of Lewis Acid catalysts can be considered: $AlCl_3$, HF/BF_3, $FeCl_3$, CF_3SO_3H. In using HF/BF_3 as combined catalyst and solvent, Dale [117] prepared PEK of inherent viscosity (η_{inh}) from 0.8 to 1.65 with useful processability and toughness. The MW was controlled through the use of biphenyl or *p*-phenoxy benzophenone as chain terminators.

The use of HF/BF_3 or CF_3SO_3H for high MW polymers is costly and requires special equipment with significant precautions for either catalyst system.

Aluminum chloride in methylene chloride solvent with a corresponding Lewis Base such as DMF, LiCl, or NH_4Cl results in a complex which can be a solvent or a swelling agent for the polymer. The ensuing polymers exhibit good melt stability because the polymeric system is more responsive to end-capping agents due to the Lewis Base.

Recently a poly ether ketone ketone, known as PEKK, was commercialized [118]. PEKK is prepared with a controlled amount of branching without any adverse effect on polymer properties (Eq. 29).

$$\text{Ph–O–Ph} + \text{ClCO–C}_6\text{H}_4\text{–COCl} \xrightarrow[\text{PhNO}_2]{\text{AlCl}_3} \left[\text{–C}_6\text{H}_4\text{–O–C}_6\text{H}_4\text{–CO–C}_6\text{H}_4\text{–CO–} \right]_n$$

Equation 29

A comparison of PEEK and PEKK indicates that PEKK is about 10 °C higher in T_g, slightly higher tensile strength but lower in toughness (Table 9). PEKK has a lower melt viscosity than PEEK.

The lower melt viscosity greatly facilitates melt impregnation of carbon fibers. Unidirectional AS4 with 58% Vf results in excellent mechanical properties due to optimum consolidation of melt impregnated tows and high matrix resin modulus.

McGrath [119] has shown that the attractive properties of PEK can be used in an entirely different mode of composite fabrication. Rather than utilize high MW PEK and its excessively high melt viscosity for transforming into composites, he has prepared low MW diamine terminated PEK (Eq. 30) oligomers (reaction 1). On heating (reaction 2) further reaction or crosslinking occurs via ketimine formation and confirmed by solid state ^{13}C NMR analysis. Results to date have shown that readily processable, tough and solvent resistant crosslink PEK type materials can be prepared with fracture toughness values similar to polycarbonates.

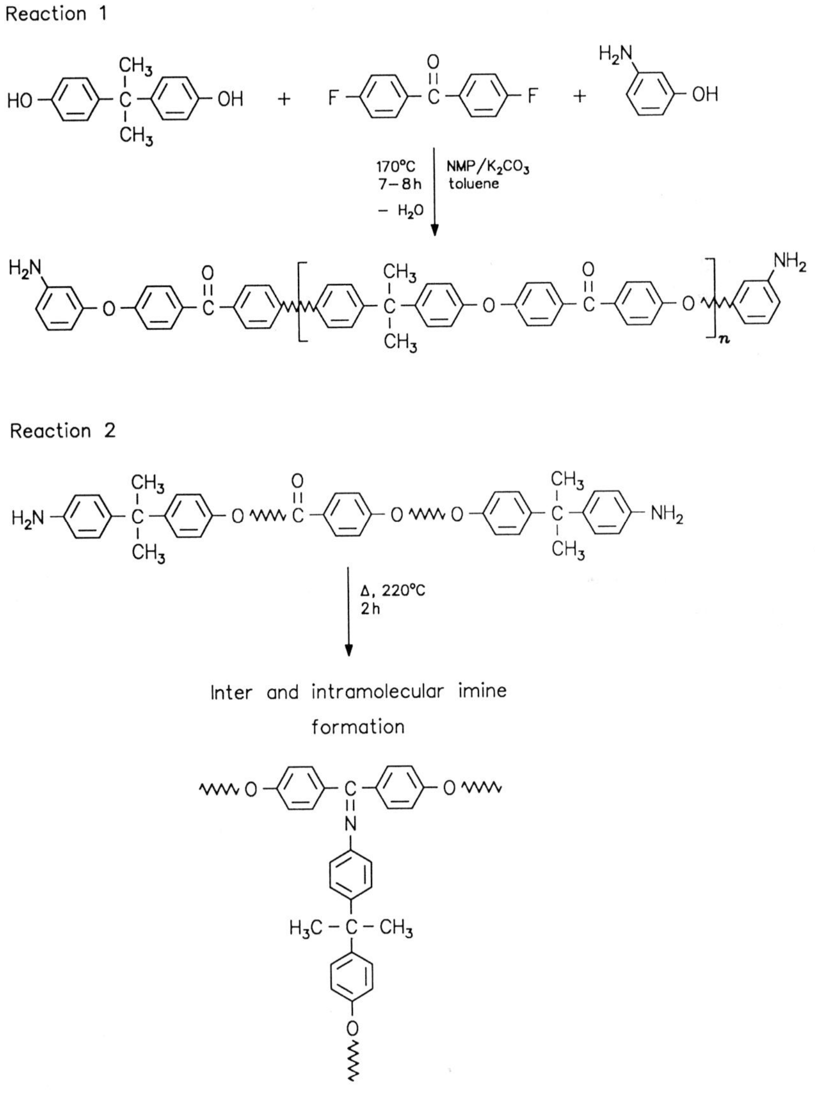

Reaction 1

CH_3 HO – C – OH + F – C(=O) – F + H_2N – OH
CH_3

170°C
7–8h
– H_2O
NMP/K_2CO_3
toluene

H_2N – O – C(=O) – [– C(CH_3)(CH_3) – O – C(=O) – O]$_n$ – NH_2

Reaction 2

H_2N – C(CH_3)(CH_3) – O ∿∿ C(=O) – O ∿∿ O – C(CH_3)(CH_3) – NH_2

Δ, 220°C
2h

Inter and intramolecular imine formation

∿∿ O – C – O ∿∿
‖
N
|
H_3C – C – CH_3
|
O
∿∿

Equation 30

2.4.2 Thermoplastic Polyimides (TPI)

It was mentioned previously in the polyimide section that there are some polyimides that are thermoplastic and can be processed as high performance thermoplastic resins. The objective of developing TPI is to improve processability of the polyimide without sacrificing high temperature characteristics. These resins include LARC-TPI, NR-150B2, Polyimide 2080 and Amide–Imide resins.

(1) LARC-TPI

This system based on benzophenone dicarboxylic dianhydride and 3,3′ diamino benzophenone was developed by NASA in the late 1970s [120] (Eq. 31). Under a licensing agreement, Mitsui Toatsu Chemical (Japan) has prepared large quantities of LARC-TPI for evaluation in space and military applications. The resin system is available from MTC in the form of amic acid solutions, film, powder or prepreg. Rogers (US), also a licensee of the NASA technology, has identified their material as Durimid. Processability of LARC-TPI appears to be related to the use of a meta substituted diamine with a flexible linkage (C=O) between the benzene rings. A variety of different linkages for the diamine and the dianhydride was examined in the preparation of different types of LARC-TPI resins [121]. As large quantities of LARC-TPI became available, some anomalous resin behavior was observed for the powdered imidized resin. St. Clair and coworkers [122] found that upon rheological characterization, the material exhibited a very low melt viscosity that continued to increase and was related to a form of transient crystallinity. St. Clair prepared a higher MW version of imidized powder. In this study the level and type of crystallinity can vary and relate to cyclodehydration method. Levels of crystallinity were between 40 and 50% for η_{inh} of 0.3 while the MTC powder is about 45% crystalline with η_{inh} of 0.15.

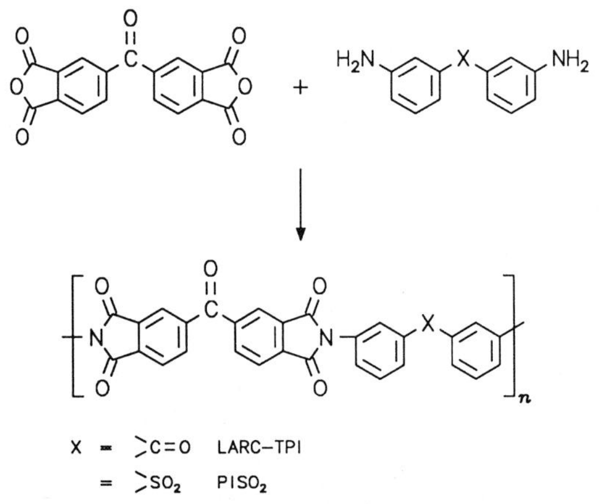

X = >C=O LARC–TPI

= >SO_2 $PISO_2$

Equation 31

The other commercially available material from Rogers is known as Durimid [123]. A comparison of mechanical properties of MTC and Durimid type LARC-TPI resin is shown in Table 9.

Durimid has higher strength and modulus with a twofold increase in impact.

If 3,3′ diamino diphenyl sulfone is reacted with 3,3′,4,4′ benzophenone tetracarboxylic dianhydride, a product known as $PISO_2$ is obtained [124] (Eq. 31). When LARC-TPI is blended (see polymer blends) with $PISO_2$ improved processability of both components occurs.

Closely related studies by Hergenrother [125] that combine LARC-TPI technology with aryl ether ketone functionality have resulted in new semi-crystalline polyimides known as LARC-CPI. The general reaction scheme consists of the reaction of an aryl dicarboxylic dianhydride with new diamines containing carbonyl and ether connecting groups between aromatic rings (Eq. 32):

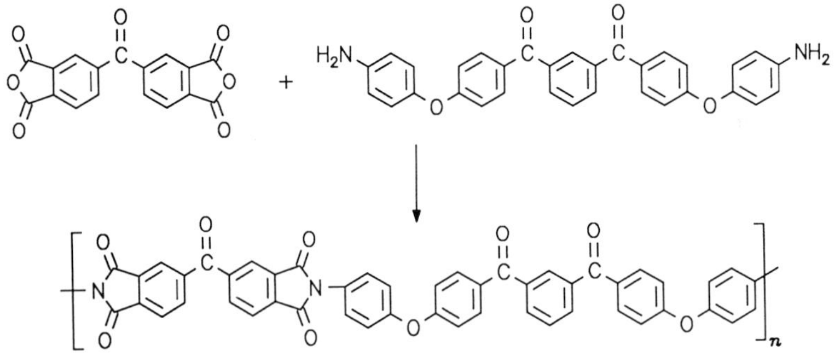

LARC–CPI

Equation 32

The LARC-CPI (Table 9) exhibits a high modulus and fracture toughness and maintains 90% of tensile strength and modulus after four days at 316 °C in air.

(2) NR-150 B2

NR-150 B2 is an amorphous linear thermoplastic polyimide with outstanding high temperature properties ($T_g = 350\,°C$). The resin is prepared by reacting hexafluoro isopropylidene dianhydride with 95/5 molar ratio of *p*/*m* phenylene-diamine [126] (Eq. 33):

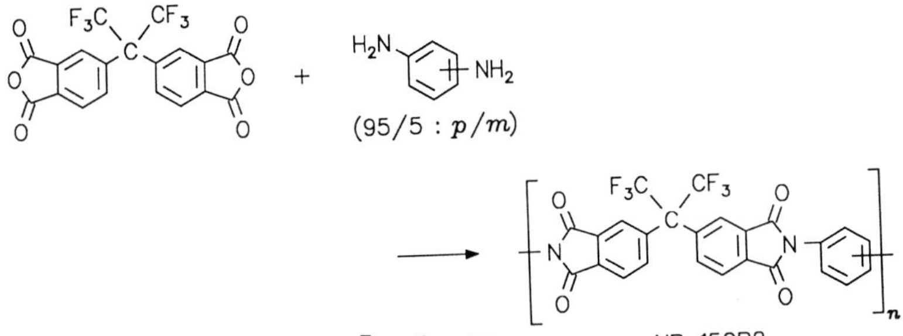

(95/5 : *p*/*m*)

Equation 33

NR 150B2

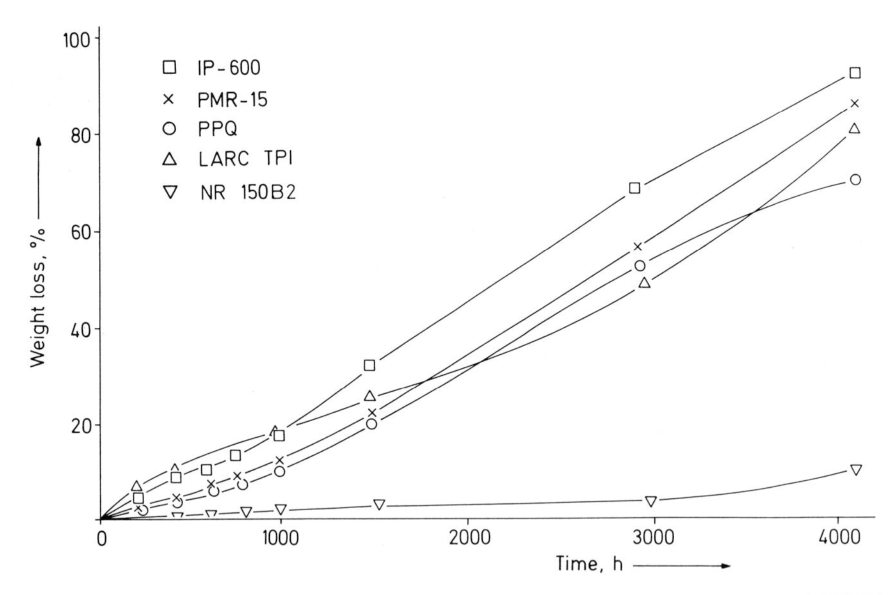

Fig. 3. Thermooxidative stability of high temperature polymers; temperature 316 °C in flowing air, 100 cm·min^{-1}

This TPI resin system is remarkable (Table 9) in that it possesses strength, stiffness, toughness, and unusually high T_g with long term thermooxidative stability. (It should be noted that NR-150 B2 was combined with PMR-15 in the API section to yield LARC RP40.) A comparison of long term isothermal thermooxidative stability of various polyimides (condensation, addition and thermoplastic) by Scola [127] identified NR-150 B2 as the most stable polyimide (Fig. 3). Long term isothermal thermooxidative stability was determined by measuring changes in weight in flowing air. After more than 4100 hours NR-150 B2 suffered only about a 10% weight loss. A distant second place was shared by PMR-15 and poly (Phenyl-Quinoxaline) (XXX).

XXX

A fiber reinforced composition based on NR-150 B2 and chopped carbon fibers has been introduced by Dupont and is known as Avimid N [128]. Studies of fabricated complex shapes indicate that Avimid N offers potential for use in jet engine components as well as other uses where temperatures in excess of 340 °C are required.

Another thermoplastic amorphous polyimide [129] known as Avimid K-III (Dupont) is approximately 100 °C lower in T_g than NR 150 B2 or 251 °C but can be used continuously to 232 °C. Avimid K-III demonstrates excellent strength, toughness, damage and environmental resistance. It possesses extraordinary toughness with G_{IC} of 19 kJ/m^2. Boeing Military Airplane Company (BMAC) has chosen Avimid K-III for use in the fabrication of the aft fuselage and wings of the BMAC prototype of the Advanced Tactical Fighter (ATF).

(3) Polyetherimide

The Polyetherimide resin system combines the high heat resistance of the imide groups with the ductility of the aryl ether component resulting in a high performance amorphous thermoplastic material. It is commercially available as Ultem from General Electric [130, 131]. It is prepared by the nucleophilic displacement of a dinitro bisimide by a disodio bisphenol (Eq. 28). The aromatic imide provides rigidity, creep resistance and high heat resistance while the aryl ether is responsible for good melt processability and flow due to chain flexibility (Table 9). Ultem has poor solvent resistance, being susceptible to attack by hydraulic fluids and common aircraft solvents.

A CF prepreg version of Ultem known as CYPAC X7005 (American Cyanamid) is reported to resist common solvents without adversely affecting processing and performance characteristics of Ultem [132]. Its toughness is impressive with a G_{IC} of 23 kJ/m^2. Blends (see blends) can be prepared when polyetherimide is added to polyamide–imide type polymers [133, 134] or PAEK resins [135].

(4) Polyimide 2080

A somewhat unconventional method is used to prepare Polyimide 2080. It consists of the reaction of benzophenone dicarboxylic acid dianhydride with a mixture of 80/20 toluene diisocyanate and methylene 4,4′ diisocyanate (Eq. 34):

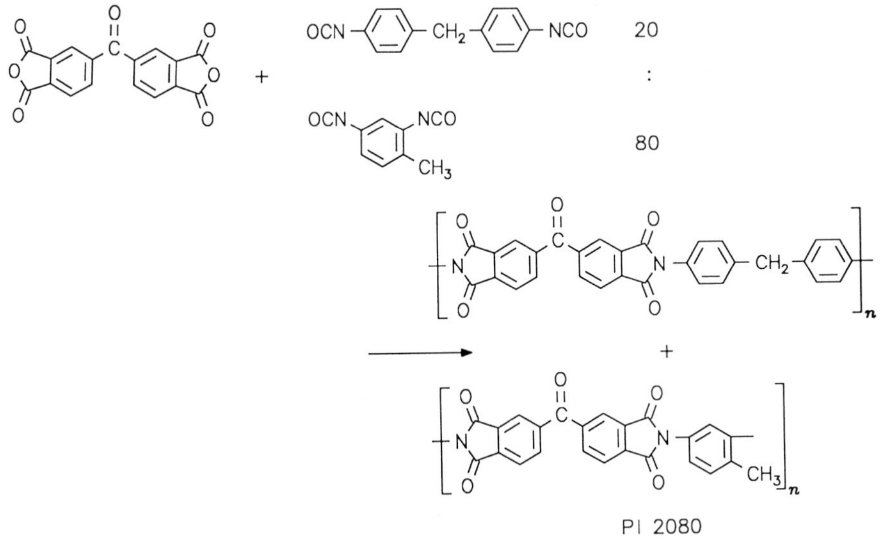

OCN–C$_6$H$_4$–CH$_2$–C$_6$H$_4$–NCO 20 : 80 OCN, NCO, CH$_3$

CH$_2$ n + CH$_3$ n

PI 2080

Equation 34

Although PI 2080 is a moderately low MW polymer, it is a moldable material and quite different to the typical intractable products that are obtained from the reaction of aryl diamines with aromatic dianhydrides (Table 9).

An unusually high T_g blend ($>300\,^{\circ}C$) of 35% PI 2080 and 65% BMI/MDA is reported by Yamamoto [136]. Glass and carbon fiber composites with this blend maintained their mechanical properties up to 260 °C.

(5) Amide–Imide

Analogous to polyetherimide, the polyamide–imide (PAI) is another polymeric system which possesses both the high heat resistance of the imide group as well as the toughness and ductility of the amide functionality. The commercially available PAI is known as Amoco's Torlon and is prepared by reacting trimellitoyl acid chloride with a mixture of 70/30 4,4′ diamino diphenyl oxide and phenylene diamine [137] (Eq. 35):

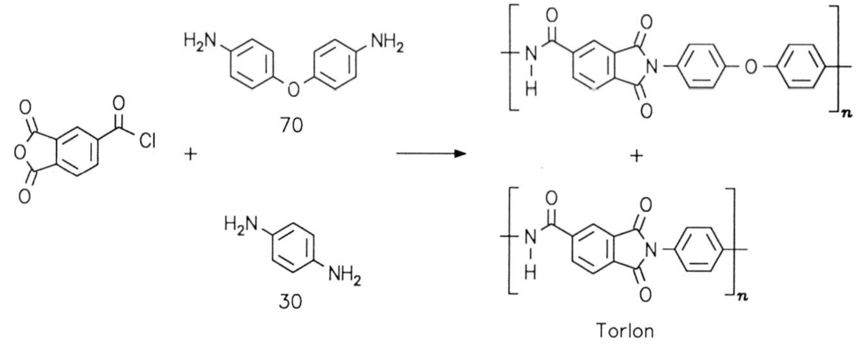

Torlon

Equation 35

Torlon is an amorphous TP but requires post cure for MW increase and ultimate T_g value for the resulting matrix resin. Some of the key features of Torlon are its high strength or high transverse and shear strength in composites combined with excellent toughness and damage tolerance (Table 9). It also exhibits good adhesion to fiber and this relates to overall beneficial composite properties.

New improved polyamide–imide resins have been reported with improved flow and total cure. With these new improvements, original properties of PAI have been retained [138]. In spite of its amorphous characteristics, it exhibits the performance of a semi-crystalline material with excellent solvent resistance.

A family of fluoro containing PAI resins [138a] with improved processability, reduced moisture sensitivity and chemical resistance has been recently reported. Depending on whether MDA or oxy-dianiline is used as the diamine, materials with T_gs above Torlon are obtained (280–300 °C). The polymers can be melt processed into thin films or thick pressings. Moisture sorption comparison of Torlon and fluorinated PAI indicate that 50% less moisture is absorbed by the fluoro PAI.

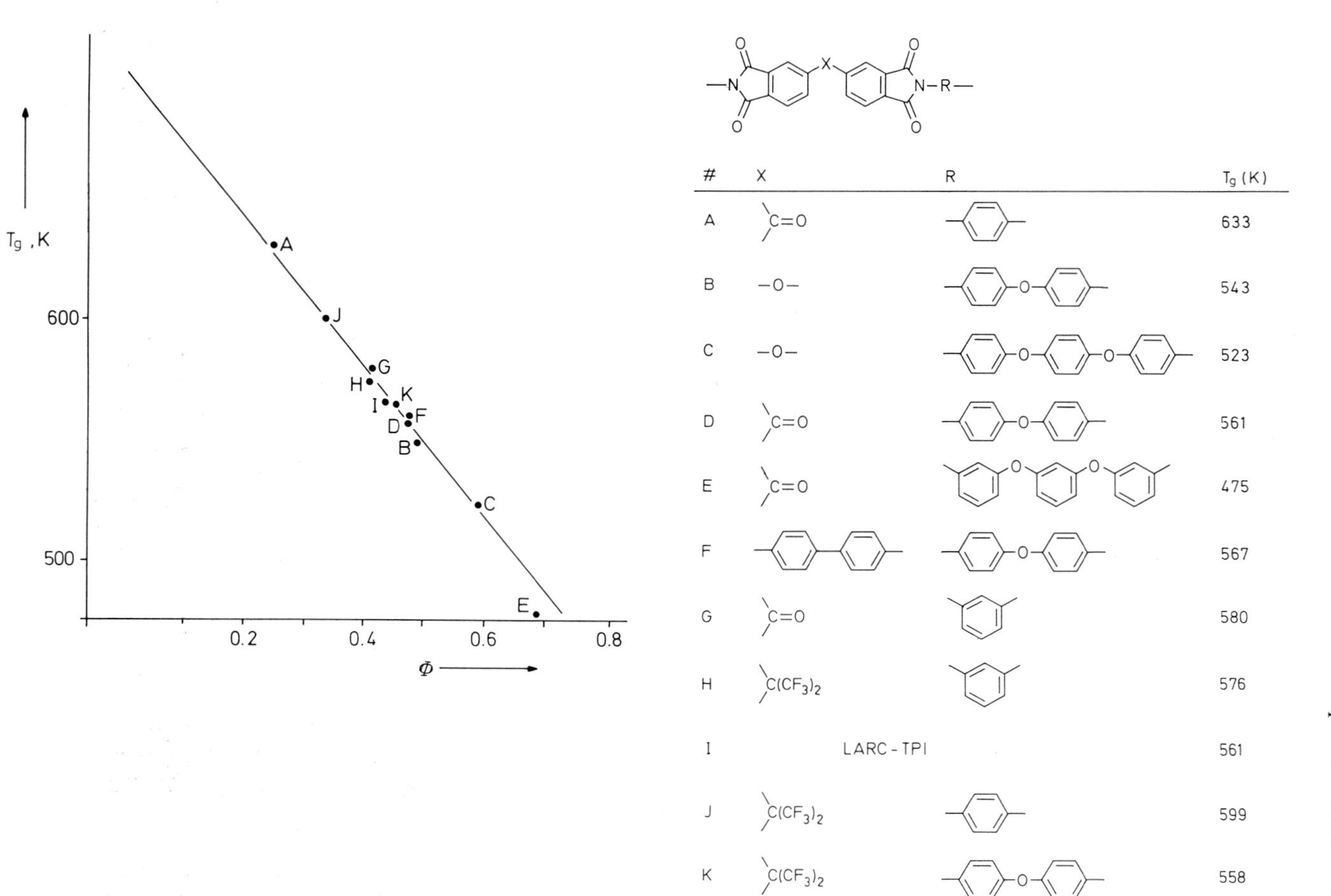

Fig. 4. Relationship between T_g and ether linkage density for polyimides

2.4.2.1 T_g Correlation

Lee [139] has developed a correlation between T_gs of many semi-rigid backbone structures such as polyimides, polyquinoline and polyquinazolines and ether equivalences. The latter are obtained by quantum chemical calculations for rotational barrier energies as well as group contributions to heat capacity jumps that are obtained from experimental measurements. Over 250 polymers were examined in developing this T_g correlation [139]. Figure 4 is a plot of this correlation.

The correlation developed by Lee allows one to compute T_g values of unknown or yet to be prepared polyimides.

2.4.3 Polyarylene Sulfide

The most prominent polyarylene sulfide is polyphenylene sulfide (PPS). It is a semi-crystalline resin with high temperature stability, inherent flame resistance and excellent char resistance. Although PPS type materials have been known for almost 100 years, its commercial production (Phillips) began in 1973 [140] (Eq. 36). PPS is obtained as an off-white powder. It is a linear polymer with a MW from 16000 to 35000 Dalton (Table 9). Other commercial processes for PPS have been reported by Kureha (Japan) and Bayer (Europe).

$$Cl-C_6H_4-Cl + Na_2S \xrightarrow{\text{polar organic solvent}} [-C_6H_4-S-]_n + 2\,NaCl$$

PPS

Equation 36

PPS can be transformed into a tougher material by thermal treatment. Heating it to its melting point (or 285 °C) in air involves chain extension, branching and ultimately crosslinking. In spite of its low T_g, PPS maintains reasonable strength and possesses a heat deflection temperature of 263 °C at 1.82 MPa that is only about 10 °C lower than Torlon.

Closely related materials based on similar reaction conditions have been reported [141] and consist of polyphenylene sulfide ketone (PPSK) and polyphenylene sulfide sulfone (PPSS) (Eq. 37):

$$Cl-C_6H_4-X-C_6H_4-Cl \xrightarrow{NaSH,\ Na_2S,\ NMP} [-C_6H_4-X-C_6H_4-S-]_n$$

X = >C=O; PPSK

>SO₂; PPSS

Equation 37

Newer PPS amorphous systems known as PAS-1 and PAS-2 with higher T_gs have been reported by Phillips [142, 143]. These resins have high temperature mechanical performance, good mechanical strength, and good solvent resistance.

2.4.4 Summary of Thermoplastic Resin Performance and Future Prospects

Thermoplastic resins continue to display exemplary ductility/toughness for improved damage tolerance, high temperature resistance and strength. Most TP prepregs are dry, rigid, solvent-free prepregs with no tack. It is anticipated that composite fabrication will avoid expensive and costly autoclave cycles by using cost efficient processing methods such as compression molding and will substantially reduce fabrication cycles from the customary 24 hour autoclave cycle to about 1 hour. The resulting TP composites will be more durable, less easily damaged and potentially more easily repaired. Nevertheless long term temperature versus creep characteristics need to be further defined as well as solvent/chemical resistance. Solvent resistance will become more critical since many amorphous TP resins are being promoted as viable TP resin systems and appear to be displacing some of the early semi-crystalline TP resins [144]. It is possible that solvent/chemical resistance may require a post-surface treatment of the composite (e.g. plasma, crosslinking, etc.) to achieve solvent/chemical resistance and possibly minimize creep.

2.5 Ordered Molecules

A unique class of high performance polymeric materials has evolved over a period of several decades. These consist of macromolecules which are highly ordered or aligned. Some of these materials are known as liquid crystalline polymers and are characterized by a molecular architecture that results from the assembly of rigid, rod-like or "ordered" molecules [144a]. These materials are ordered (crystalline) in both solid and liquid phases. These liquid crystalline polymers (LCP) in solution are *lyotropic* or in the melt are *thermotropic* and are between the boundaries of solid crystals and isotropic liquids. Flory [145] predicted that polymeric materials in solution exhibit liquid crystalline characteristics due to comparatively elongated, rigid-rod molecules. Commercial LCPs consist of aromatic rings connected by either amide linkages (Nomex, Kevlar) or ester groups (Xydar, Vectra).

The highly innovative efforts of Morgan and Kwolek at Dupont [144a, 146, 147] led to the development of aromatic polyamides and in particular Nomex and Kevlar fiber. The latter is a lyotropic (solution processed) polymer. Kevlar is discussed in more detail in the fiber section.

The Dupont team demonstrated that high strength, stiffness and stability can be attained within the polymeric material through a judicious control of molecular architecture. By using aromatic rings linked with amide groups to maintain a rigid molecular alignment, polymers of very high strength and stiffness were obtained. In solution these rigid rod polymers associate themselves in highly ordered domains or form liquid crystals and become further orientated when drawn into fibers. These LCP are lyotropic since they are solution formed into films or fibers but are *not* melt processable. The use of flexible links between aromatic rings to yield LCPs which are thermotropic and melt processable were due to the efforts of Economy while at Carborundum [148], Jackson of Eastman [149], and the Celanese team of Calundann and Jaffe [150]. The area of thermotropic LCPs was recently reviewed by Chung [151].

Economy, in 1973, developed aromatic polyesters which were thermotropic LCP but difficult to process. The Carborundum LCP technology was purchased by Dart and Kraft resulting in the introduction of Xydar (XXXI) in 1984. Later Amoco acquired the Xydar products from Dart. The monomeric units that comprise Xydar are *p*-phenyl biphenol (PPBP), *p*-hydroxy benzoic acid (HBA) and terephthalic acid (TPA). A closely related thermotropic LCP developed by Celanese is Vectra (XXXII) which is based on TPA, HBA and hydroxy naphthanoic acid (HNA).

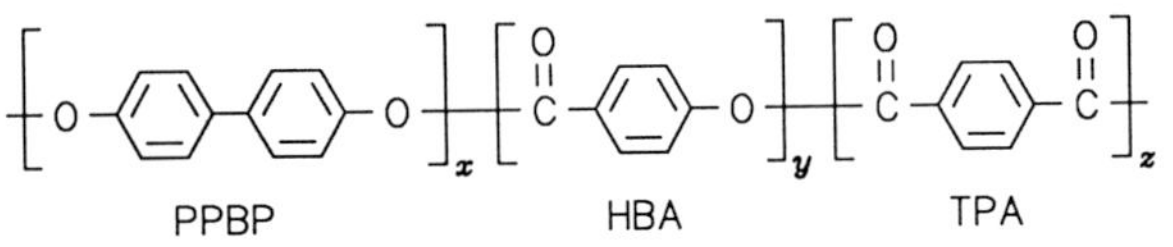

PPBP HBA TPA

XYDAR

XXXI

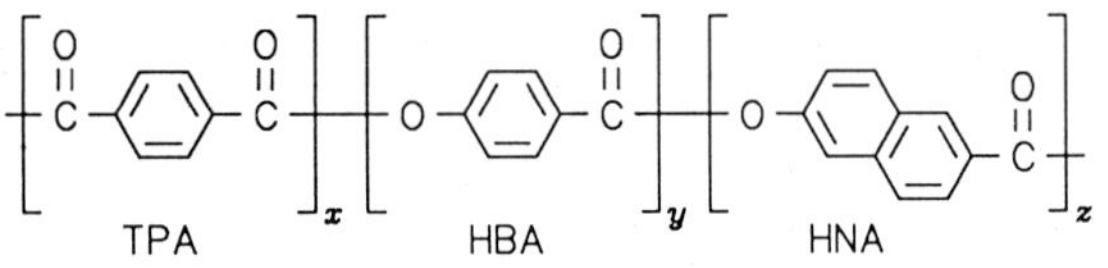

TPA HBA HNA

VECTRA

XXXII

Properties of Xydar and Vectra are listed in Table 10. The attractive HDT of Xydar of 355 °C is slightly lower than Vespel (Dupont Polyimide) HDT of 360 °C. However Vespel is not injection moldable. Xydar is UV- and γ-radiation tolerant with a high degree of crystallinity, a typical characteristic of LCP. Xydar is virtually inert to all corrosive chemicals and solvents, especially those encountered in electronics, chemical processing, automotive or aerospace applications. Although, Vectra has a lower HDT than Xydar, Vectra possesses higher strength and modulus (Table 10).

Other rigid rod or ordered polymers are composed of heterocyclic aromatic components which are the main repeat unit of the macromolecule [152]. The

Table 10. Properties of Xydar and Vectra

		Xydar (SRT-300)	Vectra (C130)
T.S.	MPa	116	165
T.M.	GPa	9.5	15.6
Elong. %		4.9	2.0
Flex St.	MPa	131	219
Flex mod.	GPa	14.3	14.3
Notched izod	J/m	144	132
Heat deflection °C	1.82 MPa	355°	240 °C
Continuous temp. use	(°C)	240	200

use of heterocyclic aromatic components allows multiple bond linkages for these rigid rod polymers when compared to the previously mentioned LCPs that are linked by a single bond (e.g., amide, ester). A rigid rod heterocyclic aromatic system that is commercially available (Hoechst Celanese) as either a fiber or a powder is polybenzimidazole (PBI). It is prepared by the reaction of tetraamino biphenyl and diphenyl isophthalate to PBI (Eq. 38):

$$(H_2N)_2C_6H_3-C_6H_3(NH_2)_2 + Ph-O-\overset{O}{\overset{\|}{C}}-C_6H_4-\overset{O}{\overset{\|}{C}}-O-Ph \longrightarrow \text{PBI}$$

Equation 38

Polybenzimidazole powder is known as Celazole which can be transformed into molded parts by a proprietary sintering process [153].

Properties of Celazole are listed in Table 11. Celazole PBI is an unique polymeric material which is non-combustible in air and exhibits exceptionally high temperature resistance and solvent resistance. It is further characterized by an unusually high compressive strength. Materials molded of Celazole can survive short excursions to temperatures as high as 740 °C. Celazole's unique property characteristics lend it to many aerospace, military and electronic applications which require premium polymer properties.

Miscible blends (see blends/alloys) of PBI with a variety of thermoplastic polyimides such as PI 2080, Ultem 1000, or LARC-TPI result in polymer miscibility, single T_g systems and clear films [154]. These factors strongly support polymer compatibility. The authors proposed compatibility or miscibil-

Table 11. Properties of Celazole

		(U-60)
Density	kg/m^3	1.3×10^3
T.S.	MPa	160
T.M.	GPa	5.9
Elong	(%)	3
Flex St.	MPa	220
Flex mod	GPA	6.5
Compressive strength		
at yield (12% strain)	MPa	400
at 10 % strain	MPa	340
Notched izod	J/m	30
T_g (DMA)	°C	425

ity is due to intermolecular interactions involving N-H of PBI with C=O groups of the polyimide. Steiner [155] has shown that PBI can be blended with bismaleimide resins to form a tougher IPN type system. (See IPN section).

An extensive research program sponsored by the U.S. Air Force in conjunction with Stanford Research Institute has identified several rigid rod or highly ordered polymers which consist of heterocyclic aromatic units. The most prominent are polybenzothiazole (PBT) [156] and polybenzoxazole (PBO) [157]:

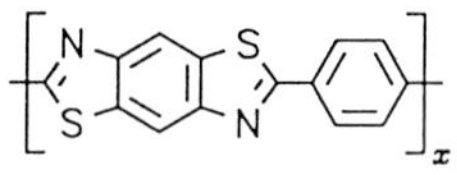

PBT 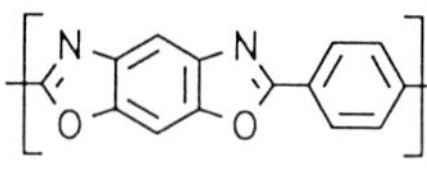 PBO

Less rigid heterocyclic systems related to PBT and PBO have also been prepared [158] and are ABPBI [159], ABPBT and ABPBO:

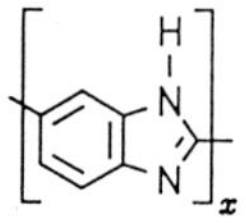

ABPBI 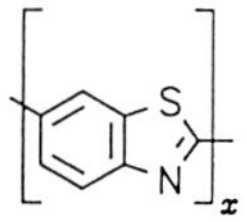 ABPBT 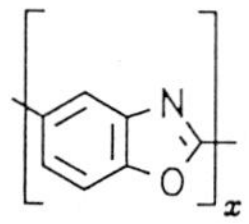 ABPBO

Recently Dowell [160, 161] predicted the existence of super-strong liquid crystalline polymers (SSLCPs). These SSLCPs are proposed to be the first polymers to have excellent compressive strength as well as tensile strength, and modulus greater than existing strong LCPs along with good processing characteristics. These specially designed SSLCPs have LCP backbones consisting of aramid, PBT or PBO with side chains of similar or related LCPs which align themselves in interdigitated structures, Figure 5.

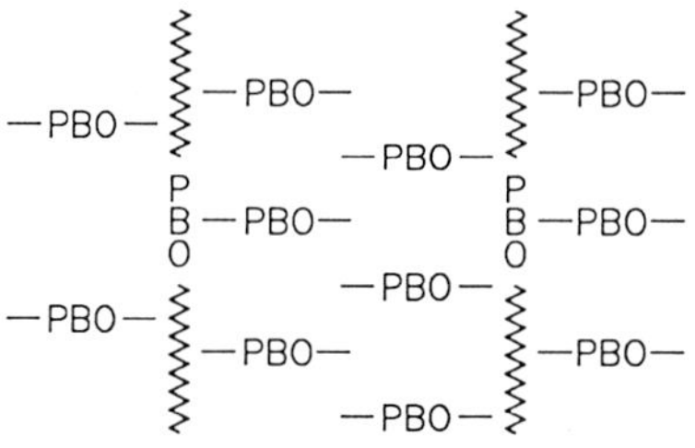

SS LCP

Fig. 5

Table 12. Comparison of mechanical properties of SSLCPs and LCPs

	T.M (GPa)	T.S. (GPa)	Compressive strength (GPa)
LCPs	130–365	3.5–5.8	0.4–0.5
SSLCPs	600–945	9–12	9–12

Calculated increases [161] in mechanical properties of several SSLCPs are compared with properties of existing LCPs (Table 12). Synthesis of these SSLCPs to confirm mechanical property predictions are in progress at Los Alamos National Laboratories.

2.6 Molecular Composites

During the last two decades, Helminiak of the Air Force Wright Aeronautics Laboratories and Wolfe at SRI have been developing novel aromatic heterocyclic polymers for high performance applications as well as a new challenging area known as "molecular composites."

Experimental fibers (see fiber section) of PBT and PBO have exhibited excellent mechanical properties with very high specific strengths and moduli. However these materials like most organic fibers (Kevlar, Spectra) lack compressive strength.

One of the more innovative features of these rigid rod systems has been the attempt to transform these materials into molecular composites. A molecular composite is viewed as a rigid rod polymer molecularly dispersed in a random coil polymer matrix. Presently a macroscopic composite contains a rigid rod as reinforcing agent or fiber imbedded within a random coil matrix resin. Due to the high aspect ratio (l/d) of the rigid rod polymer, a higher modulus and greater strength is anticipated for the molecular composite. Other proposed advantages of a molecular composite consist of little or no interface problems or lack of adhesion between matrix and reinforcing fiber since reinforcement is at a molecular level. Further no separate handling of either fiber or matrix is necessary avoiding expensive and time consuming operations (weaving, prepregging, etc.).

Several strategies have been pursued in the preparation of molecular composites. These include (1) molecular entanglement, (2) graft copolymer or in situ polymerization and (3) block copolymer method. The two latter techniques provide methods to chemically attach the rigid rod molecules and the random coil matrix molecules together in a block or graft (in situ) polymerization copolymer.

2.6.1 Molecular Entanglement

This technique requires careful processing of solutions of both components to prevent an undesirable aggregation and phase separation of the rigid-rod polymer. Helminiak and coworkers [162] determined the necessary entanglement MW for matrix flexible coil polymers and MW of rigid rod reinforcement for molecular composites. Entanglement of PBT and nylon 66 with both materials possessing Mn 35 000 led to a 30/70 rigid to flexible composition that behaved like a polymer solution "rigidized" by the rigid rod component while compositions varying from 100/0, 70/30 and 50/50 exhibited flexible coil behavior. Helminiak [163] has recently demonstrated a novel process for fabricating bulk articles of PBT/Nylon 66 molecular composites. The process involves the consolidation of coagulated, wet molecular composites (PBT/N66) into three dimensional shapes for mechanical property determinations. The authors estimated from mechanical properties that the reinforcement efficiency of PBT was 50–70% of maximum value.

Takayanagi [164] has examined aramid reinforcing polymers as a rigid rod polymer with flexible aliphatic polyamides such as Nylon 6 and 66. The resulting molecular composites by coagulation led to higher modulus for the composite due to the presence of aramid microfibrils with diameters of 10–30 nm. It is suggested that a facile transfer of the aramid modulus occurred in the composite resulting in higher overall polymer modulus.

2.6.2 Graft Copolymer/In situ Polymerization

Some of the early work of in situ molecular composite polymerization was conducted by Tsutsui and Tanaka [165] who formed poly-L-glutamic acid rods with polyethylene oxide chains. Mathias and Moore [166] similarly used aramid as the rigid rod substrate and attached either nylon 3 or epoxy flexible polymer groups to PPD-T by in situ polymerization (Eq. 39):

NaH
DMSO

Acrylamide or
Epoxy/Amine

Kevlar

N3 or Epoxy

rigid

flexible

Equation 39

Table 13. Mechanical properties of PPD-T/N3

% PPD-T	T.S. (MPa)	T.M. (GPa)	Elong (%)
—	28	1.06	5.2
10	60	3.15	3.0
30	19	17.55	9.4

Table 14. Mechanical properties of epoxy-based composite materials

System	Tensile strength (MPa)	Tensile modulus (GPa)	Elongation (%)
Epoxy/MPDA (control)	42 ± 10	2.0 ± .2	2.4 ± .5
Epoxy MPDA 10% pulp (macroscopic composite)	58 ± 7	2.1 ± .3	3.9 ± .5
Epoxy/MPDA 10% PPTA anion	81 ± 5	2.3 ± .1	4.9 ± .3

Films of aramid/N3 exhibited greatly improved strength and modulus with no loss in flexibility when compared to unmodified N3 (Table 13).

With epoxy as the flexible coil [167], the epoxy segment was appended to aramid pulp, a highly fibrillated form of Kevlar. In Table 14, properties of epoxy/MPDA control, a macroscopic composite of the 2 components and the molecular composite composed of Kevlar/epoxy are compared. Higher strength, higher modulus and elongation are observed with the Kevlar/epoxy molecular composite.

2.6.3 Block Copolymer

In 1980 Takayanagi [168] had shown that PPD-T block copolymer with nylon 66 was superior to a blend of similar materials. Preston [169] reported significantly improved properties for block copolymers of flexible amide hydrazides with rigid polybenzamide.

Arnold and coworkers [170] prepared an ABA block copolymer of benzimidazole (ABPBI) and PBT that increased the C_{cr}, critical concentration solution (C_{cr} = concentration at which isotropic to anisotropic phase transition occurs), characteristics of the block copolymer as well as tensile strength when they are compared to a macroscopic blend of both components. The triblock copolymer (XXXIII) is structurally: 30%PBT/70%ABPBI.

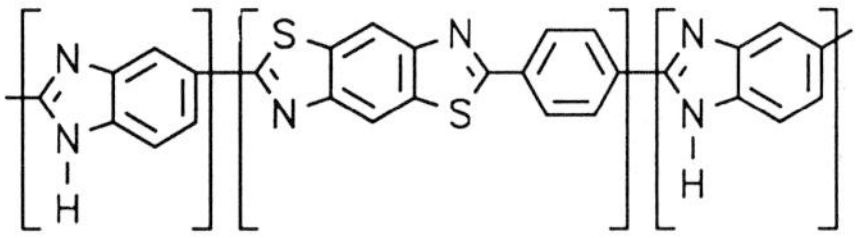

ABPBI PBT ABPBI

XXXIII

A careful examination of the morphology and mechanical properties of the PBI/ABPBI block copolymer has been conducted by Krause [171]. Mechanical properties of PBT, ABPBI and block copolymer fibers are summarized in Table 15.

When copolymer ($C < C_{cr}$) fiber was spun, no large scale phase separation was observed. WAXS analyses indicated that both ABPBI and PBT crystallites were present in a much smaller size. Transmission Electron Microscopy dark field images suggested that a dispersion of PBT in ABPBI at a level of 3 nm or greater had been achieved. Krause claims that no crystalline phase separation can occur at a scale larger than 3 nm. Thus a copolymer coil/rod/coil with this level of dispersion is considered to be a molecular composite. The efficient reinforcement of the matrix at a molecular level and the orientation of ABPBI and PBT would be expected to yield high mechanical properties for the block copolymer spun fiber. However, SEM micrography of the copolymer fiber indicated a void content of 30 to 40% in the fiber interior suggesting that the mechanical properties of the copolymer (Table 15) can be further optimized with a void-free fiber.

Table 15. Mechanical properties of PBT, ABPBI and block copolymer fibers

Fiber	Tensile modulus (GPa)	Tensile strength (MPa)	Elongation (%)
PBT	320	3100	1.1
ABPBI	36	1100	5.2
Block copolymer ($C < C_{cr}$)[a]	100	1700	2.4

[a] Solution concentration below critical concentration at which isotropic to anisotropic phase transition occurs

Optimum structures with molecular composite properties are still the goal as one evaluates the results to date of the three techniques. Continued progress may indeed achieve the key objectives of molecular composites such as improved processing economics and a lower density composition when molecular composites are compared with the customary macroscopic continuous filament matrix composite route. If molecular composites succeed, it is possible these unique materials may be injection moldable and become formidable competition to thermoplastic high performance engineering polymers as well as labor intensive advance composite materials.

2.7 Interpenetrating Networks

The incorporation of a thermoplastic segment or resin within a crosslinked system or combining a thermoplastic component with a thermoset material results, in some instances, in an enhanced improvement in the properties of the final system. Molecular entanglement of two moderately incompatible or in some cases partly compatible materials is a characteristic feature of an interpenetrating *network* (IPN) or a semi-IPN system [172, 173].

Usually no chemical bond formation occurs between the two different resins or oligomer systems. The dissimilar polymers are entangled or entrapped in such a manner that no phase separation occurs. A further characteristic of the IPN is the distribution of stress and rigidity within the dissimilar phases. Several different methods can be used to prepare IPNs:

1. A preformed crosslinked system is dispersed in a monomer which swells the network by penetrating into network interstices. Heat and catalyst cures the monomer.
2. Two different monomers or prepolymers are heat/catalyst cured. Each monomer or prepolymer will homopolymerize and not co-react with each other.

2.7.1 Semi-Interpenetrating Networks

A Semi-IPN (S-IPN) consists of a single component crosslinking while the other is or will form a linear chain. Usually a thermoplastic resin is dissolved in the reactive monomer or oligomer. Upon curing the monomer or oligomer, an S-IPN is obtained. Analogously, a linear thermoplastic resin can be synthesized in the presence of a dispersed crosslinked resin.

Semi-IPN materials based on thermoplastic and thermoset components have become quite popular within the last decade. Semi-IPN popularity is attributable to property enhancement of both dissimilar phases. Thermoplastic properties deficiencies such as solvent, abrasion and heat resistance are significantly improved by the thermoset phase while the ductility of the thermoplastic segment improves the limited elongation or brittleness of the thermoset component.

Semi-IPN systems are well documented in high performance compositions such as Bisphenol A dicyanates which crosslink in the presence of linear thermoplastic polymers (polycarbonate, polysulfone or polyetherimide [174]). The resulting S-IPN exhibited greater strength and toughness relative to a conventional thermoset system emerging from crosslinked dicyanate. Superior processing and heat resistant properties relative to any of the thermoplastic resins were also noted.

Yamamoto [175] obtained an S-IPN by reacting BMI in the presence of the thermoplastic polyimide 2080. The most desirable composition consisted of 35% 2080 and 65% BMI. Tougher, high strength composites with T_g ($>300\,^\circ C$) were obtained.

Increased toughness of modified BMI systems occurred when Stenzenberger [176] introduced thermoplastic resins such as polysulfone, polyetherimide or polyhydantoin. Heat resistance of S-IPNs with either polysulfone or polyetherimide was reduced due to lower T_gs of the thermoplastic resins. However the heat resistancc S-IPN with polyhydantoin was unaffected with compositions containing as much as 33% polyhydantoin. Toughness of these S-IPN systems carried over to carbon fiber laminates.

Steiner [177] also observed improved toughness of a modified BMI when polybenzimidazole was dispersed into BMI formation. Although exceptional toughness was obtained when BMI/PBI components were combined, only marginal toughness was observed for the corresponding carbon fiber composite.

Epoxy matrix resins have been modified into S-IPNs with selected high performance polyaryl ethers by Sefton and coworkers [178]. Two S-IPN systems were developed for composite aircraft primary structures. Morphological characterization of these systems indicated that one S-IPN exhibited a two phase spinodal morphology while the other was phase inverted with the epoxy phase discontinuous and the thermoplastic being continuous. Both S-IPNs displayed significantly improved toughness in both neat resin and laminate properties without compromising other mechanical properties.

Previously it was mentioned that PMR-15 is quite brittle, and composites containing PMR-15 as matrix resin are susceptible to microcracking. Through the S-IPN technique, a tougher and more microcrack resistant PMR-15 resin can be obtained [179]. Combining thermoplastic polyimide resin, NR-150 B2, with PMR-15, NASA Langley researchers developed a S-IPN known as LARC-RP40. It exhibited a significant improvement in resin fracture toughness and composite microcracking resistance (over PMR-15) while maintaining a high T_g of 369 °C, processability, mechanical properties and thermooxidative stability.

Most S-IPN are composed of dissimilar components. St. Clair [180] was able to prepare an S-IPN with identical repeat units in the heat curable acetylene terminated polyimidesulfone ($ATPISO_2$) and polyimidesulfone ($PISO_2$):

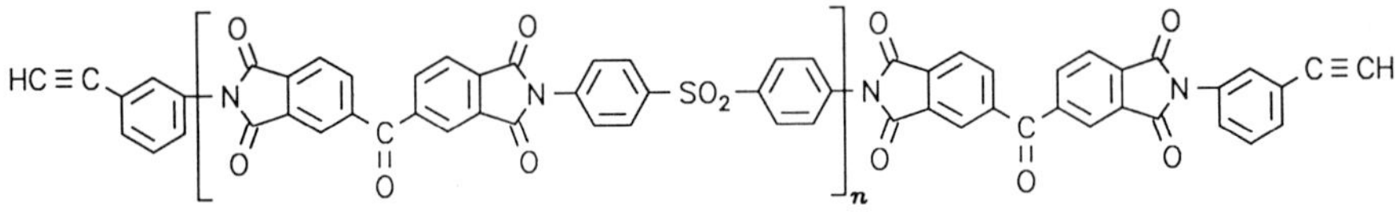

ATPISO$_2$

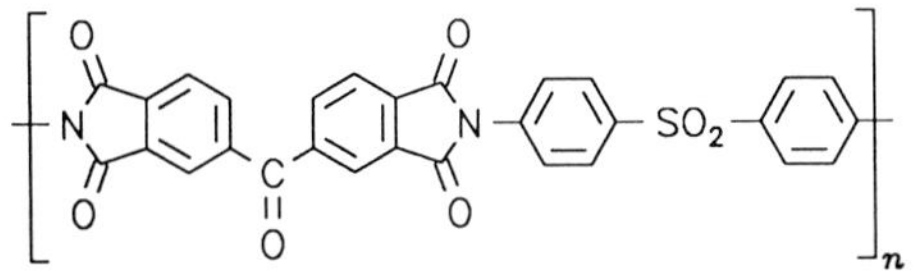

PISO$_2$

Table 16. T$_g$s of ATPISO$_2$/PISO$_2$/S-IPN

Component	T_g (°C)	Wgt loss (300 hours, isothermal)
PISO$_2$	275	linear loss with increased temp.
S-IPN	302	2%
ATPISO$_2$	330	4%

The unique S-IPN displayed a T_g intermediate between the T_g of PISO$_2$ and the T_g of thermoset ATPISO$_2$ (Table 16). Based on torsional braid analysis (TBA) the authors proposed that a single phase exists for the S-IPN as evidenced by a narrow damping peak. The combination of similar repeat units for the S-IPN of polyimidesulfone facilitated the processing of the linear thermoplastic component (PISO$_2$) and contributed improved thermooxidative stability.

These examples of S-IPN provide sufficient evidence that further improvements and more S-IPN systems will be developed in the future. It is further anticipated that these newer S-IPN systems will display favorable characteristics such as: ease of processing during prepregging or filament winding; satisfactory fiber adhesion; tack and drape; ease of curing; little or no microcracking; and balanced composite mechanical properties for fracture toughness.

2.8 Alloys or Blends

The mixing of two polymeric materials can result in a single phase, compatible system known as an alloy or in a 2 phase composition known as a blend [181]. Alloys or blends differ from IPN or semi-IPN materials since only mechanical mixing of the two resins is required for preparation. Comparison of these binary compositions is as follows:

	Alloy	Blend
Phase	Single, compatible	2-phases
Compatibility	Full	Little or none
T_g	Single	Two T_gs
Film appearance	Clear	Clear or opaque

Few single phase, miscible alloys exist; the majority of the polymer binary mixtures are two phase or blend systems.

An alloy which has been well characterized is the PPO/PS or Noryl system. The Noryl family of products continue to be a commercial success and is a stimulus to polymer scientists to discover other high performance alloy products. Blends usually offer properties that are often an average of the component polymers. In some cases a small amount of a third resinous component is added and provides some compatibility between the two incompatible resins.

Small angle neutron scattering (SANS) is used to determine phase separation of molecules and measure phase separated microstructures up to 1 μm. SANS provides some insight in the understanding of phase separation mechanisms and blend strength. It can also identify polymer blends that cannot be directly observed. Information generated by SANS is quite valuable especially when blends are examined in the molten state. It can determine whether a possible irreversible separation between both components can occur at elevated temperatures and whether the thermoplastic blend can be injection moldable (Chapter 4, Analysis/Testing).

Recently alloys of aromatic polybenzimidazoles (PBI) and aromatic polyamides have been reported by Karasz and coworkers [182]. According to the authors, polymers of these generic types appear to be miscible over a wide range of compositions and structural variations. Compositions of PBI with a variety of polyimides such as polyetherimide (PEI) and. thermoplastic polyimides like LARC-TPI and PI 2080 (see thermoplastic polyimide section) exhibited single composition dependent T_gs and well defined tan relaxations associated with these T_g values. The resulting clear films and enhanced solvent resistance further corroborate the miscibility of PBI with the various imide resins. Based on FTIR analyses the authors claim that the miscibility between PBI and PI 2080 is due to intermolecular interactions between the N–H of PBI and carbonyl group of PI 2080.

Alloys/Blends with PEI

Polyetherimide demonstrates a unique capability of alloy or blend formation with a variety of high performance polyaryl ethers.

The alloying characteristics of PEI are documented by the studies of Robeson [183] and Brugel [184]. Robeson found that polyaryletherketones, prepared by nucleophilic conditions, will yield alloy systems with PEI. Compositions varying from 70/30 to 30/70 PAEK/PEI exhibited a single, sharp T_g, improved toughness and better solvent resistance for alloys containing PEI.

Similarly work by Brugel identified a polyetherketone via Friedel–Crafts conditions, that was readily transformed into a single T_g, compatible system. Depending on the weight ratio of PEK TO PEI, the T_g of the alloys varied from 159 °C (100% PEK) to 186 °C (50/50 composition) to 218° (100% PEI).

A somewhat different procedure has been reported by Holub [185]. Solutions of intermediate amic acid imide of both polyamide imide and PEI were mixed and films were cast at about 300 °C to yield an abrasion resistant film. T_gs were generally below the T_g of either component such as polyamide imide (241 °C) or PEI (218 °C) and suggest that further interchange occurs between polymer intermediates prior to final imide reaction.

A melt blending procedure [186] of PEI with polyamide imide yielded a 2-phase blend with separate T_gs being observed for each component in the blend.

2.9 References

1. J. K. Gillham, Makromol. Chem. Macromol. Symp. 7, 67 (1987); M. T. Aronhime and J. K. Gillham, "Advances in Polymer Science," Vol. 78, K. Dusek, editor, pp. 83–113, Springer, Berlin, Heidelberg New York, 1986.
2. S. Sherman, J. Gannon, G. Buchi and W. R. Howell, "Epoxy Resins" Encyclopedia of Chemical Technology, Vol 9, p. 267+ (1980) Wiley, New York.
3. L. V. McAdams, J. A. Gannon, "Epoxy Resins" Encyclopedia of Poly. Sci. and Eng. Vol 6. pp. 322–382; H. Mark, editor, 1986, John Wiley, New York.
4. "Epoxy Resins, Chemistry and Technology," C. A. May, editor, Marcel Dekker, New York 1988.
5. K. Hodd, "Epoxy Resins," Comprehensive Polymer Science, Vol 5, p. 667+. G. Allen and J. C. Bevington, editors, Pergamon, Oxford, 1989.
6. R. S. Bauer SAMPE 34, 1889 (1989).
7. H. Lee and K. Neville, "Handbook of Epoxy Resins," McGraw-Hill, New York, 1967.
8. T. Kamon and H. Furukawa, "Advances in Polymer Science," Vol 80, K. Dusek, editor, Springer, Berlin Heidelberg New York, 1986.
9. R. S. Bauer, A. G. Filippov, L. M. Schlaudt and W. T. Breitigam, SAMPE 32, 1104 (1987); SAMPE 33, 1385 (1988).
10. P. Delvigs, Poly Comp. 7(2), 101 (1986).
11. M. Fischer and R. Schmid, Colloid & Polymer Sci., 264, 387 (1986).
12. A. R. Siebert, Makromol. Chem. Macromol. Symp. 7, 115 (1987).
13. J. McGrath et al., SAMPE 32, 1276 (1987).
14. H. C. Gardner, M. J. Michno, Jr., G. L. Brode and R. J. Cotter (UCC) U.S. 4,661,559 (4/28/87).
15. D. K. Kohli and M. M. Fisher (Amer. Cyan.) U.S. 4,645,803 (2/24/87).
16. R. S. Bauer and H. D. Stenzenberger, SAMPE 34, 899 (1989).
17. G. R. Alman et al., SAMPE, 33, 979 (1988); SAMPE 34, 259 (1989).

18. H. G. Recker et al., SAMPE, 34, 747 (1989); 21st Int'l Tech. Conf. SAMPE 21, 283 (1989)).
19. C. B. Bucknall and A. H. Gilbert, Polymer 30, 213 (1989).
20. A. Knop, V. Böhmer, Louis A. Pilato, "Phenol-Formaldehyde Polymers," Chapter 35, Volume 5, in "Comprehensive Polymer Science," Sir Geoffrey Allen and J. C. Bevington, editors, Pergamon, Oxford, 1989.
21. A. Knop and L. A. Pilato, "Phenolic Resins," Springer, Berlin Heidelberg New York, 1985.
22. P. W. Kopf, Encyclopedia of Polymer Science and Engineering, Vol II, p. 45, H. Mark et al. editors, John Wiley, New York, 1988.
23. Rogers Corp. Manchester, CT 06040.
24. V. L. Landi and B. B. Fitts, U.S. 4,659,758 (4/21/87) and 4,725,650 (2/16/88) Rogers.
25. Temephen D50, phenolic dispersion resin available from Temecon, Piscataway, N.J. 08854.
26. J. P. Critchley, G. J. Knight and W. W. Wright, "Heat Resistant Polymers," p. 29, Plenum, New York, 1983.
27. P. V. Mosher U.S. 4,585,837 (April 29, 1986); U.S. 4,650,840 (March 17, 1987) Hitco.
28. Plastics Engineering Co., Sheboygan, Wisc. 53082.
29. M. K. Gupta, D. W. Hoch, G. Salee, U.S. 4,785,040 (11/15/88) Occidental Chemical Corp.
30. B. M. Culbertson, O. Tiba, M. L. Deviney and T. A. Tufts, SAMPE 34, 2483 (1989).
31. B. Amram and F. Laval, J. Appl. Poly. Sci., 37(1), 1 (1989).
32. I. Serlin, E. Lavin and A. H. Markhart, Chapter 37, "Handbook of Adhesives," 2nd Edition, I. Skeist, editor, Van Nostrand, N.Y. 1977.
33. P. G. Cassidy and N. C. Fawcett, "Encyclopedia of Chemical Technology," 18, p. 704 (1982) Wiley, NY.
34. J. W. Verbicky, Jr. "Encyclopedia of Polymer Science and Engineering," Vol 12, 364, H. Mark et al. editors, John Wiley, NY, 1988.
35. B. Sillion, Polyimides and other Heteroaromatic Polymers, "Comprehensive Polymer Science," Volume 5, p. 499, G. Allen and J. C. Bevington, editors, Pergamon, Oxford, U.K. 1989.
36. R. F. Sutton, Jr. U.S. 4,472,153 (1988) (Dupont).
37. H. R. Lubowitz, U.S. 3,528,950 (1970).
38. T. T. Serafini, P. Delvigs and G. Lightsey, J. Appl. Poly. Sci., 1b, 905 (1972); U.S. 3,745,149 (1973) NASA.
39. Composites, Vol 1, Engineered Materials Handbook, p. 801 + , ASM International U.S. 1987.
40. "Advanced Composite Materials," D. J. DeRenzo, Noyes Data Corp., Park Ridge, NJ 1988.
41. Composites and Laminates, Vol 1, D.A.T.A. Inc., San Diego, CA 1987.
42. "High Temperature Polymer Matrix Composites," T. T. Serafini, editor, Noyes Data Corp., Park Ridge, NJ 1987.
43. J. Roberts and R. D. Vannucci, SAMPE J. Mar/April 24 (1986).
44. R. E. A. Escott, 19th Int'l SAMPE Tech. Conf. 19, 398 (1987).
45. L. J. Sullivan and R. Ghaffarian, SAMPE 33, 1604 (1988).
46. D. Wilson, Brit. Poly. J., 20, 405 (1988).
47. Dexter Hysol, Cleveland, Ohio.
48. C. H. Sheppard SAMPE Quart. (Jan), 14, (1987).
49. R. D. Vanucci, SAMPE 32, 602 (1987).
50. R. H. Pater and C. D. Morgan, SAMPE J., 24(5), 25 (1988).
51. T. L. St. Clair and R. Jewell, SAMPE 23, 520 (1978).
52. T. L. St. Clair and D. J. Progar, SAMPE 24(2), 1081 (1979).
53. L. J. Baldwin, M. B. Meador and M. A. Meador, Poly Preprints 29(1), 236 (1988).
54. W. Gorham, J. Poly. Sci. PT. A-1, 4, 3027 (1966).

55. N. Bilow, A. L. Landis and L. J. Miller, U.S. 3,845,018 (1974).
56. A. L. Landis, N. Bilow, R. H. Boschan, R. E. Lawrence and T. J. Aponji, Poly. Preprints 15(2), 537 (1974).
57. National Starch, Bridgewater, N.J. 08807.
58. D. J. Capo and J. E. Schoenberg, SAMPE J. Mar–Apr, 35 (1987).
59. P. M. Hergenrother, Poly Preprints 27(2), 408 (1986).
60. P. M. Hergenrother, J. Poly. Sci. Polym. Chem. Ed. 20, 3131 (1982).
61. F. I. Hurwitz, L. H. Hyatt, L. D'Amore, H. X. Nguyen and H. Ishida, Polymer, 29, 184 (1988).
62. A. Renner, U.S. 4,667,003 (1987); A. Renner and S. H. Eldin, U.S. 4,666,997 (1987) Ciba-Geigy.
63. M. A. Chaudhari, B. Lee, J. King and A. Renner, SAMPE 32, 597 (1987).
64. Plastics World, January 1988, p. 64.
65. F. E. Arnold et al., Poly. Prep 26(2), 176, 178 (1985).
66. F. E. Arnold, SAMPE 31, 968 (1986).
67. F. E. Arnold, J. Poly. Sci. Part A, Poly Chem 25(11), 3159 (1987).
68. M. P. Cava and M. J. Mitchell, "Cyclobutadiene and Related Compounds," Academic NY, 1967, Chapter 6.
69. L. R. Denny and E. J. Soloski, Poly. Prep. 29(1), 194 (1988).
70. F. E. Arnold et al., SAMPE 33, 240 (1988).
71. F. E. Arnold, SAMPE 31, 968 (1986).
72. L-S Tan, F. E. Arnold and E. J. Soloski, J. Poly. Sci. Part A, Poly Chem. 26, 3103 (1988).
73. H. D. Stenzenberger, Chapter 2. Polymer Matrices in "Carbon Fibres and Their Composites," E. Fitzer, editor, Springer, Berlin Heidelberg New York, 1985.
74. Y. Ohnuma, T. Sugimoto, Y. Nemoto and K. Kanayama, SAMPE, 32, 33 (1987).
75. G. T. Kwiatkowski, L. M. Robeson, G. L. Brode and A. W. Bedwin, J. Poly. Sci. Poly. Chem. Ed. 13, 961 (1975).
76. J. E. McGrath et al., Poly. Prep. 27(1), 315 (1986); ibid, 28(1), 77 (1987); ibid, 29(1), 346 (1988); SAMPE 33, 1080 (1988).
77. I. K. Varma and S. Sharma, J. Comp. Mat. 20, 308 (1986).
78. S. J. Shaw and A. J. Kinlock, Int. J. Adh 5(3), 123 (1985).
79. S. Takeda and H. Kakiuchi, J. Appl. Poly. Sci. 35, 1351 (1988).
80. K. F. M. G. J. Scholle and H. Winter, SAMPE 33, 1109 (1988).
81. Advanced Composites, July/August p. 45 + (1988).
82. H. D. Stenzenberger et al., SAMPE 31, 920 (1986); ibid, 32, 44 (1987); Brit. Poly. J. 20(5), 383 (1988).
83. H. D. Stenzenberger, SAMPE 33, 1546 (1988).
84. M. S. Sefton et al., 19th Int'l SAMPE Tech. Conf. 19, 700 (1987).
85. D. A. Shimp et al., 18th Int'l SAMPE Tech. Conf. 18, 851 (1986).
86. Y. Yamamoto, SAMPE 30, 903 (1983).
87. Rhone Poulenc, Louisville, KY.
88. E. P. Woo and D. J. Murray, U.S. 4,713,442 (12/87) Dow; D. A. Jarvie, SAMPE, 33, 1405 (1988).
89. S. Das and D. C. Prevorsek, U.S. 4,831,086 (5/89) Allied-Signal.
90. S. Das, D. C. Prevorsek and B. T. DeBona, 21st Int'l SAMPE Tech. Conf. 21, 972 (1989).
91. D. A. Scola, Polyimide Resins in Composites, Vol. 1 Engineered Materials Handbook, ASM International, Metals Park, Ohio, 1987.
92. J. T. Gotro et al., Polymer Composites, 8, 39 (1987).
93. D. A. Shimp et al., SAMPE 33, 754 (1988).
94. E. S. Hsiue and R. L. Miller, SAMPE 30, 1035 (1985).
95. D. A. Shimp et al., 18th Internal SAMPE Tech. Conf. 18, 851 (1986).
96. T. P. Hefner, U.S. 4,680,378 (7/14/87); 4,683,276 (7/28/87) Dow.

97. R. N. Johnson and A. Farnham, J. Poly. Sci. A-1, 5, 2415 (1967); R. N. Johnson, in "Encyclopedia of Polymer Science and Technology," editors, N. Bikales, H. F. Mark and N. G. Gaylord, Wiley, N.Y. 1969, Vol. 11, p. 447+.
98. S. Maiti and B. K. Mandel. Prog. Poly Sci, 12, 111 (1986).
99. M. J. Mullin and E. P. Woo, J. Macro. Sci. Rev. Macromol. Chem. Phy. C27(2), 313 (1987).
100. P. M. Hergenrother, B. J. Jensen and S. J. Havens, Polymer, 29, 358 (1988).
101. C. P. Smith, Chem. Tech. 290, May 1988.
102. Comprehensive Polymer Science, Vol 5, Chapters 29 and 33, G. Allen and J. C. Bevington, editors, Pergamon, Oxford, 1989.
103. B. E. Jennings, M. E. B. Jones and J. B. Rose, J. Poly. Sci. Part C 16, Part 1, 715 (1967).
104. J. E. McGrath et al., Poly. Prep. 25(2), 19 (1984).
105. J. B. Rose, British 1,414,424 (1975) I.C.I.
106. J. B. Rose and P. A. Staniland, U.S. 4,320,224 (1982).
107. T. Nagurmo, H. Nakamura, Y. Yoshida and K. Hiraoka, SAMPE 32, 396 (1986).
108. D. C. Leach, F. N. Cogswell and E. Nield, SAMPE 31, 434 (1986).
109. G. T. Spaner and N. O. Brink, SAMPE 33, 284 (1988).
110. N. J. Johnston and P. M. Hergenrother, SAMPE 32, 1400 (1987).
111. F. N. Cogswell et al., SAMPE 32, 382 (1987).
112. D. K. Mohanty, S. D. Wu and J. E. McGrath, Poly. Prep. 29(1), 352 (1988).
113. D. M. White et al., J. Poly. Sci. Polym. Chem. Ed 19, 1635 (1981).
114. T. Takekoshi, Polymer J. 19, 191 (1987).
115. D. E. Florigan and I. W. Serfaty, Modern Plastics. June, p. 146 (1982).
116. W. H. Bonner, U.S. 3,065,205 (1962).
117. W. Dale, U.S. 3,956,240 (1976).
118. E. G. Brugel, U.S. 4,720,537 (1988); I. Y. Chang, SAMPE Quarterly, 19(4), 29 (1988).
119. J. S. Senger, D. K. Mohanty, C. D. Smith and J. E. McGrath, Poly. Prep. 29(1), 358 (1988).
120. V. L. Bell, J. Poly. Sci. Poly. Chem. Ed., 14, 2275 (1976).
121. A. Yamaguchi and M. Ohta, SAMPE J. Jan/Feb 28 (1987).
122. T. L. St. Clair et al., SAMPE Quart. 18(1), 1(1986); ibid. 18(4), 20 (1988).
123. D. C. Sherman, C-Y Chen and J. L. Cercena, SAMPE 33, 134 (1988).
124. N. J. Johnston and T. L. St. Clair, SAMPE J. Jan/Feb 12 (1987).
125. P. M. Hergenrother and S. J. Havens, SAMPE July/Aug. 13 (1988); SAMPE 34, 963 (1989).
126. H. H. Gibbs, J. Appl. Poly. Sci., Appl. Poly. Sym. 5, 207 (1979).
127. D. A. Scola and J. H. Vontell, Chem Tech. Feb. 112 (1989).
128. H. H. Gibbs, SAMPE 33, 1473 (1988).
129. R. J. Boyce, T. P. Gannett, H. H. Gibbs and A. R. Wedgewood, SAMPE 32, 169 (1987).
130. D. M. White et al., J. Poly. Sci., Poly Chem. Ed. 19, 1635 (1981).
131. J. W. Verbicky, "Polyimides" Encyclopedia of Poly. Sci. and Eng. Vol. 12, p. 364 and Mark, Bikales, Overberger, Menges, editors, 1988, Wiley-Interscience NY.
132. S. L. Peake, A. Maranci and E. Sturm, SAMPE 32, 420 (1987).
133. F. F. Holub and G. A. Mellinger, US 4,258,155 (3/24/81). (G. E.).
134. G. T. Brooks, U.S. 4,640,944 (2/3/87) (Standard Oil, Indiana).
135. J. Harris and L. M. Robeson, J. Appl. Poly. Sci. 35(7), 1877 (1988).
136. Y. Yamamoto, S. Satoh and S. Etoh, SAMPE 30, 903 (1985).
137. B. Cole, SAMPE 30, 799 (1985).
138. Plastics Technology p. 27, Dec. 1988.
138a. M. J. Jaffe et al., 21st Int'l SAMPE Tech. Conf. 21, 755 (1989).
139. C. J. Lee, SAMPE 34, 929 (1989).

140. J. F. Geibel and R. W. Campbell, "Poly(phenylene sulfide)s" Chapter 32, in Comprehensive Polymer Science, Sir G. Allen and J. C. Bevington, editors, Vol. 5, Pergamon, Oxford, 1989.
141. R. G. Gaughan, U.S. 4,716,212 (12/29/89) Phillips
142. J. E. O'Conner, W. H. Beever and J. F. Geibel, SAMPE 31, 1313 (1986).
143. S. D. Mills, 21st Int'l SAMPE Tech. Conf. 21, 744 (1989).
144. D. Leeser and B. Bannister, 21st Int'l SAMPE Tech. Conf. 21, 507 (1989).
144a. S. W. Kwolek, P. W. Morgan, Encyclopedia of Polymer Science and Engineering, Vol. 9, 1 + Mark, Bikales, Overberger, Menges, editors, 1987, Wiley, NY.
145. P. Flory, Proc. R. Soc. London, A234, 73 (1956).
146. P. W. Morgan, Macromolecules, 10, 1381 (1977).
147. S. W. Kwolek, P. W. Morgan, J. R. Schaefgen and L. W. Gulrich, Macromolecules 10, 1390 (1977).
148. J. Economy, The Chemist, January 8, (1988).
149. W. J. Jackson, Jr. and H. F. Kuhfuss, J. Poly. Sci., Poly. Chem. ed. 14, 2043 (1976).
150. G. W. Calundann and M. Jaffe, "Anisotropic Polymers, Their Synthesis and Properties", Proceedings of the Robert A. Welch Conference on Chemical Research. XXVI Synthetic Polymers 1982.
151. T-S. Chung, Poly. Eng. & Sci., 26(13), 901 (1986).
152. B. Sillion, Chap. 30 in "Comprehensive Polymer Science", Sir G. Allen, J. C. Bevington, editors, Vol. 5, 1989, Pergamon, Oxford.
153. B. C. Ward, SAMPE 32, 853 (1987); ibid., 33, 146 (1988).
154. G. Guerra, S. Choe, D. J. Williams, F. E. Karasz and W. J. MacKnight, Macromolecules, 22, 231 (1988).
155. M. T. Blair, P. A. Steiner and E. N. Willis, SAMPE 33, 524 (1988).
156. J. F. Wolfe, B. H. Loo and F. E. Arnold, Macromolecules 14, 915 (1981).
157. J. F. Wolfe and F. E. Arnold, Macromolecules, 14, 909 (1981).
158. U.S. 4,533,693 (Stanford Research Institute (1985)).
159. T. E. Helminiak, Am. Soc. Org. Coat Plast. Chem. 40, 475 (1979).
160. F. Dowell, J. Chem. Phys. 91(2), 1316, 1326 (1989).
161. F. Dowell, Poly Preprints 30(2), 532 (1989).
162. D. R. Wiff, S. Timms, T. E. Helminiak and W-F. Hwang, Poly. Eng. & Sci. 27(6), 424 (1987).
163. C. S. Wang, I. J. Goldfarb and T. E. Helminiak, Polymer 29, 825 (1988).
164. F. Kumamura, T. Oona, T. Kajiyama and M. Takayanagi, Poly. Comp. 4, 141 (1983).
165. T. Tsutsui and T. Tanaka, J. Poly. Sci. Poly. Lett. Ed. 18, 17 (1980).
166. D. R. Moore and L. J. Mathias, J. Appl. Poly. Sci., 32, 6299 (1986).
167. D. R. Moore and L. J. Matthias, Poly. Comp. 9(2), 144 (1988).
168. M. Takayanagi et al., J. Macromol. Sci. B17(4), 591 (1980).
169. W. R. Krigbaum, J. Preston, A. Ciferri and Z. Shufan, J. Poly. Sci. Chem. Ed. 24, 652 (1986).
170. T. T. Tsai, F. E. Arnold and W-F. Hwang, Poly. Prep. 26(1), 144 (1985).
171. S. J. Krause, T. B. Haddock, G. E. Price and W. W. Adams, Polymers 29, 197 (1988).
172. L. H. Sperling, "Interpenetrating Polymer Networks and Related Materials", Plenum, New York, 1981.
173. L. H. Sperling, Chemtech, 104 February 1988.
174. E. S. Hsiue and R. L. Miller, SAMPE 30, 1035 (1985).
175. Y. Yamamoto, S. Satoh and S. Etoh, SAMPE 30, 903 (1985).
176. H. Stenzenberger, W. Romer, M. Herzog and P. Konig, SAMPE 33, 1546 (1988).
177. M. T. Blair, P. A. Steiner and E. N. Willis, SAMPE 33, 524 (1988).
178. G. R. Almen, R. K. Maskell, V. Malhotra, M. S. Sefton, P. T. McGrail and S. P. Wilkinson, SAMPE 33, 979 (1988).

179. R. H. Pater and C. D. Morgan, SAMPE J. 24(5), 25 (1988).
180. A. O. Egli and T. L. St. Clair, SAMPE 30, 912 (1985); U.S. 4,695,610 (9/22/87) NASA.
181. O. Oblasi, L. M. Robeson and M. T. Shaw, "Polymer–Polymer Miscibility", Academic, NY. 1979.
182. S. Stankovic, G. Guerra, D. J. Williams, F. E. Karasz and W. J. MacKnight, Polymer. Comm. 29, 14 (1988); D. J. Williams, F. E. Karasz and W. J. MacKnight, Poly. Prep. 28(1), 54 (1987); G. Guerra, S. Choe, D. J. Williams, F. E. Karasz and W. J. MacKnight, Macromolecules, 21, 231 (1988).
183. L. M. Robeson and J. E. Harris, J. Appl. Poly. Sci., 35(7), 1877 (1988); U.S. 4,713,426 (12/15/87) Amoco.
184. E. G. Brugel, U.S. 4,720,537 (Jan. 19, 1988) Dupont.
185. F. F. Holub and G. A. Mellinger, U.S. 4, 258,155 (March 24, 1981) G.E.
186. G. T. Brooks, U.S. 4,640,944 (Feb. 3, 1987) Standard Oil.

3 High Performance Fibers

3.1 Introduction

Structural materials based on ACM require strength, stiffness and toughness in conjunction with high temperature and environmental resistance. Strength and stiffness characteristics are attributable to the fiber or reinforcing agent. Only those fibers which possess very high strength and very high modulus or stiffness with moderately low density are suitable candidates as reinforcing agents in ACM. Specific modulus (normalized by fiber density) values of at least 6×10^8 cm and specific strength (normalized by fiber density) values of 6×10^6 cm are the guidelines for these high strength fibers. These include ultra high molecular weight polyethylene (UHMWPE), aramids, carbon fiber, S-2 glass, and newly reported experimental fibers, PBO and PBT. Although the specific modulus of S-2 glass is below 6×10^8 cm or only 4×10^8 cm, the specific strength of S-2 glass is 18×10^6 cm and sufficiently high to merit consideration as a high performance fiber. The high performance fibers that dominate the ACM area possess low density (950–2500 kg/m^3) and extremely high strengths (2–7 GPa) and moduli (70–800 GPa).

In Table 1 properties of high performance fibers are summarized [1, 2]. A plot of specific tensile strength and specific moduli of various high performance fibers is shown in Fig. 1.

Fibers such as PBO, T-1000 (Carbon Fiber) and Spectra 1000 (UHMWPE) are quite impressive with unusually high specific strengths with accompanying high specific moduli.

Adams [3] recently identified some common features that appear to be present in most high performance fibers:

a) very high molecular orientation
b) ordered lateral packing of molecules
c) very low concentration of axial defects

The significance of high molecular orientation relates to crystallinity and is best illustrated by the gel spinning orientation of UHMWPE into very high strength and modulus polyethylene fiber (Spectra 1000) with high crystallinity. Ordering by "liquid crystallinity" occurs when anisotropic solutions of Kevlar aramid are dry jet-wet spun into fiber. The ordering or lateral packing phenomenon is caused by lateral interactions which can be non-bonded as in UHMWPE (Spectra 1000) and PBT, hydrogen bonded as in Kevlar, or covalently bonded as graphite molecular sheets. A minimum amount of axial defects is necessary for maximum fiber strength and modulus.

Table 1. High performance fiber properties

Fiber	Density kg/m^3	Tensile strength (GPa)	Specific strength (10^6 cm)	Tensile modulus (GPa)	Specific modulus (10^8 cm)	Compressive strength (GPa)	Elongation %
UHMWPE	970	3.0	31	175	18	0.17	3.0
KEVLAR	1440	2.6	18	60–200	4.2–14	0.34–0.48	3.8
TECHNORA	1390	3.0	21	70	5.0	—	4.4
PBO	1580	5.7	36	360	23	0.2–0.4	1.9
PBT	1580	4.1	26	325	21	0.26–0.41	1.1
Pitch CF	1600–2200	0.8–2.3	5–10	38–820	2.3–38	0.48	0.25–0.50
PAN CF	1700–1900	2.3–7.1	14–37	230–830	13–26	1.05–2.75	1.5–2.4
S-2 Glass	2490	4.6	19	85	3.4	1.1	5.7

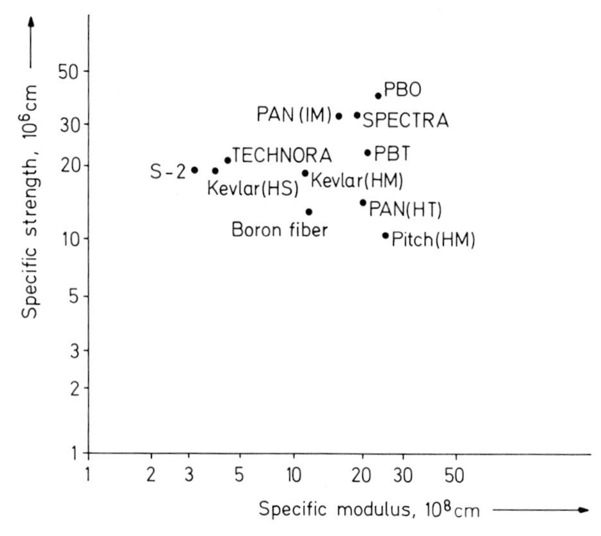

Fig. 1. Specific strength and specific modulus of high performance fibers

3.1.1 Fiber Modulus

The ultimate Young's modulus value of polymeric materials is known to be equivalent to the crystal lattice modulus in the direction of the polymer chain axis. Ultrasonic waves generated by a pulsed laser [4] have been used to measure Young's modulus of fibers. Data for PBT, carbon fiber and other high modulus fibers agreed well with those obtained by other methods. As the fiber modulus of these high performance fibers is determined, many observed moduli approach theoretical moduli values [2]. Theoretical and actual modulus values are shown in Table 2. It can be seen in Table 2 that fiber modulus is approaching 75% of theory for several fibers. With sufficient orientation and reduced defects, it is anticipated that further increases in fiber modulus will occur.

3.1.2 Fiber Strength

The corresponding determination of theoretical tensile strength of fibers is more complex. Some extrapolations of experimental tensile strength data to infinitely small fiber diameters, theoretical modulus, or simply to high draw ratios are reported in the literature. Most of the extrapolations are not justified. Several fiber effects such as molecular weight, temperature, and strain rate must be considered in determining maximum tensile strength. By attempting to develop a model based on kinetic theory of fracture and inhomogeneous distribution of stress among atomic bonds, the model developed by Tremonia and Smith [5] best described physical characteristics of fiber molecules rather than relating

Table 2. Theoretical and actual fiber modulus of high performance fibers

Fiber	Actual (GPa)	Theoretical (GPa)
Kevlar 149	200	250
UHMWPE (Spectra 1000)	175	320
PBO	360	810
PBT	325	620
Carbon fiber	490–820	1060

structure to ultimate strength. The authors' studies were applied to perfectly orientated and ordered polyethylene and poly(*p*-phenylene terepthalamide) fibers.

3.1.3 Fiber Compressive Strength

These high performance fibers are mainly utilized as reinforcing agents in Advanced Composite Materials. Many primary applications must possess sufficient compressive strength for the intended end-use application, especially if these ACM are replacing metals which exhibit good compressive strength. In high volume fraction fiber composites, compressive strength is provided by the fiber. Hence, compressive strength values of fibers are of interest and assist designers in determining approximate compressive strength of the final composite product. Compressive strengths of fibers in Table 1 are determined indirectly from unidirectional composites tested in 0° direction and then normalized to 100% fiber [2]. In most cases fiber compressive strength varies from 0.2 to slightly over 2.75 GPa with only intermediate modulus to high strength carbon fibers exhibiting high compressive strength. Organic fibers like UHMWPE, aramids, and experimental fibers, PBO and PBT, have low compressive strengths (0.2–0.43) while S-2 glass has intermediate compressive strength.

According to Kumar [2], various test methods have been developed for the determination of fiber compressive strength. Unfortunately these tests like tensile strength tests are indirect measurements of fiber compressive strength since no method exists presently that measures compressive strength of 6–15 μm diameter fibers. In the analysis of indirect methods of fiber compressive strength, other fiber or composite characteristics contribute to data scatter observed in these determinations. These include non-linear behavior in compression, fiber buckling mode in a composite, fiber misorientation, matrix resin properties and fibrillar/microfibrillar behavior. It is anticipated that refinements in fiber spinning methods, better fiber surface treatment, and minimization of fiber void content will improve fiber compressive strength.

3.2 Ultra High Molecular Weight Polyethylene

It was mentioned previously that molecular orientation, lateral packing and a low amount of axial defects were necessary to obtain fibers with high strength. Tam [6] has suggested that an additional requirement is necessary and identifies it as a high or *ultra* high MW polymer. This is especially true for high MW addition polymers such as polyethylene, polypropylene, polyvinyl alcohol and polyacrylonitrile. These polymers have been transformed into high strength fibers.

Polyethylene fibers of very high modulus and strength have been developed by hot-drawing under controlled conditions. Through the use of hot-drawing of gels or oriented crystallization of entangled networks, these techniques have been applied to linear polyethylene of very high MW ($>2 \times 10^6$). Fibers with tensile strength as high as 5 GPa and modulus to 200 GPa have been reported [7].

Recently Pennings [8] reported a PE fiber, with 7.26 GPa strength from a narrow MWD polyethylene ($M_w = 5.5 \times 10^6$ kg/kmol; Mw/Mn $\sim$ 3). This is in contrast to the early reports of Porter [9] who was able to obtain a high modulus and moderately strong polyethylene extrudate via solid state extrusion of polyethylene. Ward [10] developed a unique drawing method with a drawing temperature near the polymer melting point to yield high modulus polyethylene fibers. The Ward process has been licensed in the U.S. and Europe [6].

The preferred method of preparing UHMWPE is the gel spinning of an ultra high MW (1–5×10^6) polyethylene dissolved in a suitable solvent (xylene, paraffin oil) for untangling the polymer chains to a melt processable stage. Upon quenching the extruded fiber, the gel fiber is freed of solvent by devolatilization or extraction. Fiber drawing during or after extraction leads to high tenacity (strength) and high modulus polyethylene fiber [6]. This process or closely related gel spinning processes for UHMWPE fiber are routes to commercial PE fibers by Allied-Signal, DSM-Toyobo joint venture and Mitsui. Products available in the U.S. are Allied-Signal's Spectra 900 and 1000. Properties of UHMWPE (Spectra 1000) are contained in Table 1.

Several morphological changes occur upon transforming UHMWPE into high strength PE fiber. Processing of very large chain molecules would, by necessity, also create imperfections such as topological defects due to entanglements, twists, folds, and kinks, etc. Pennings [7] suggests that fiber formation consists of orientating an entanglement network leading to a highly oriented and highly crystalline fiber (Fig. 2). During hot-drawing of gel spun fibers a lamellar structure is preferred over an earlier proposed "shish-kebab" intermediate morphology. These morphological differences in the gel spinning process can be monitored by DSC [11] and allows one to distinguish between lamellar and shish-kebab structures of the intermediate. Optimum tensile strength of the gel spun PE fibers obtained after hot drawing depends on spinning conditions: spinning speed, temperature, spinline stretching, draw

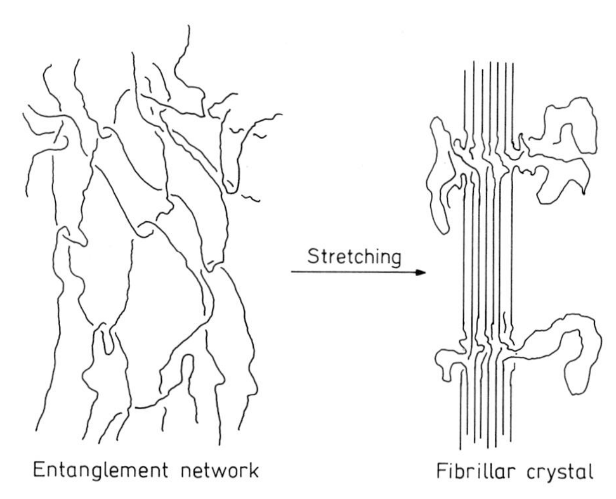

Fig. 2. Polyethylene orientation

ratio, polymer concentration and MW/MWD (Table 3). Peguy [12] has used NMR to facilitate the investigation of UHMWPE polymer entanglements consisting of thermoreversible gels of very dilute solutions [0.4 to 0.9 (w/w)].

Matsuo [13] transformed ultra high molecular weight isotactic polypropylene via the ultra drawing method to a tough film of PP with tensile strength of 1.56 GPa and 40.5 GPa modulus. This observed tensile strength of 1.56 GPa is beyond the theoretical value of 1.23 GPa reported for polypropylene fiber by Samuels [14]. This discrepancy in theoretical tensile strength magnifies the difficulties in obtaining theoretical values as well as the level of care that must be exercised in reporting early theoretical values.

Other high molecular weight addition polymers such as polyvinyl alcohol [15] and polyacrylonitrile [16] are reported to have been drawn into high strength and high modulus fibers.

Features of UHMWPE include its extremely high specific strength (31×10^6 cm), its moisture, UV, chemical and abrasion resistance, and attractive electrical properties. Unfortunately it possesses a limited upper use temperature of about 130 °C since it has a melting point of 150 °C. Both Spectra 900 and 1000 exhibit creep with Spectra 1000 having lower creep.

Many attempts have been made to improve long term properties or minimize the creep of PE fibers. Both chemical and irradiation crosslinking methods have been examined. Most crosslinking reactions after fiber drawing have led to chain scission or reduction in fiber properties. If crosslinking is conducted prior to drawing, the crosslinked material would be limited or restricted from extensive drawing because of crosslinks. UHMWPE [17] was crosslinked in the melt at 200 °C by electron beam irradiation to obtain homogeneous networks which form in non-crystalline regions. Gel–sol measurements indicated up to 100% gel was obtained with no chain scission. A change in crystalline morphology from lamellar to micellar-like crystallization occurs for the highly irradi-

Table 3. UHMWPE spinning conditions

Conc. (%)	T (°C) spin	Speed (m/min)	Winding speed (m/min)	Draw ratio	T_m (°C) (DSC)	% Cryst.	Tensile strength (GPa)
1.5	190	1	1	100	156–163	75.8	> 4.0
2.0	250	100	500	16	155–159	62.7	2.5

ated samples. Neither fiber drawing studies nor mechanical properties of crosslinked samples were reported.

The attractive high specific strength of Spectra has been responsible for its evaluation in ballistics. Its specific strength and modulus coupled to its radar transparency (very low dielectric properties) make it attractive for radome manufacture.

3.3 Aramid Fibers

The research that led to the large scale production of synthetic fibers from polyamides began around 1928 when Wallace Carothers joined Dupont from Harvard to launch a program of fundamental research. He and members of his group were responsible for the commercialization of nylon fibers in the early 1930s [18]. The continuing interest in polyamides over a period of several decades led to the development of high strength aromatic polyamides known generically as aramid fibers. Studies directed to aryl polyamides yielded a stiff chain, poly-meta-phenylene isophthalamide (MPD-I), also known as Nomex. It is an aramid with high thermal and electrical properties for apparel and electrical insulation. it is also transformed into honeycomb for strong, light weight panels.

The *para*-substituted aryl polyamide (PPD-T) forms a highly orientated fiber which is significantly stronger than MPD-I [19, 20, 21].

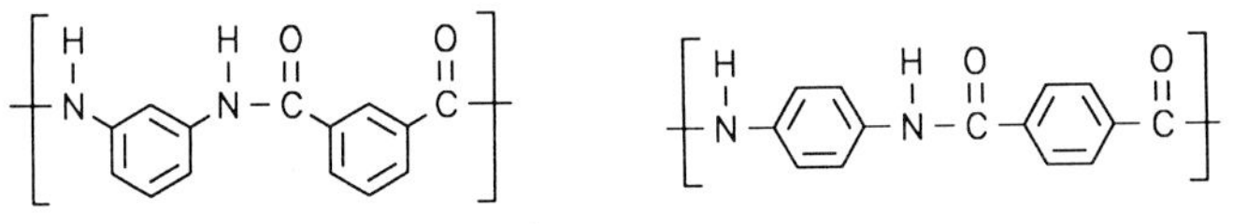

MPD–I

PPD–T

The *para*-aramid is a rigid molecule with limited flexibility adjacent to the *para* substituents of the aryl rings. Furthermore the C–N bond of the amide group has considerable double bond character which severely limits rotation. This restricted rotation contributes to the stiffness of the molecule. Interchain

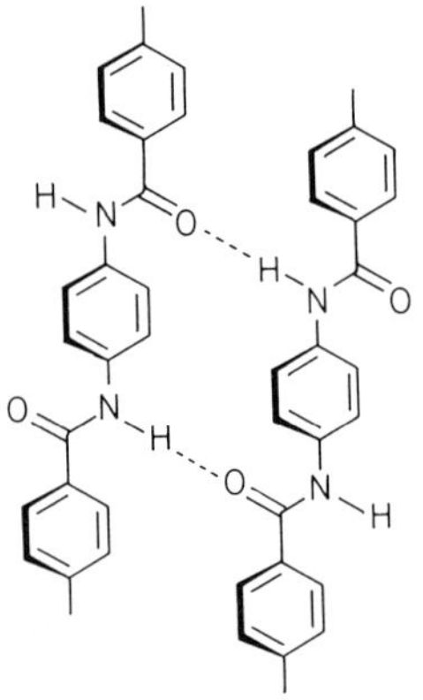

Fig. 3. Hydrogen Bonding in PPD-T

hydrogen bonding between the amide groups further enhances the rodlike configuration and morphology of PPD-T (Fig. 3).

Although single bonds are formed at the *para* positions, the molecule is sufficiently "rigidized" as is commonly observed for the multi-bonded aromatic rigid rod molecules such as PBO, or PBT.

PBO PBT

PPD-T materials in solution differ significantly from conventional flexible fiber forming materials. Dilute solutions of these rigid polymers are isotropic. As the concentration is increased, a critical concentration transforms the composition to a liquid crystalline state (Fig. 4). The anisotropy of the solution results in a corresponding reduction in viscosity.

The Mark-Houwink equation which describes the viscosity-molecular weight relationship of polymer dilute solutions is expressed as follows:

$$\eta = kM_w^\alpha$$

where η is the intrinsic viscosity, k is a constant, M_w is weight average molecular weight and α the exponent. The value of the exponent, α, varies from 0.8 for flexible polymers to 1.7 for rod-like polymers (Kevlar, PBO, PBT).

The manufacture of Kevlar fiber is as follows. The PPD-T is dissolved in concentrated sulfuric acid and dry jet-wet spun into fibers [22]. A diagram of the process is shown in Fig. 5. This method is used by Dupont and results in the family of Kevlar products [23]. Akzo (Europe) has also a PPD-T product known as Twaron and is spun from a methyl pyrrolidone-$CaCl_2$ solution. A modified PPD-T with 3,4′-diaminodiphenyl ether is currently available from Teijin (Japan) and is known as Technora.

$$\left[-NH-C_6H_4-NH-CO-C_6H_4-CO- \right]_x \left[-NH-C_6H_4-O-C_6H_4-NH-CO-C_6H_4-CO- \right]_y$$

Technora

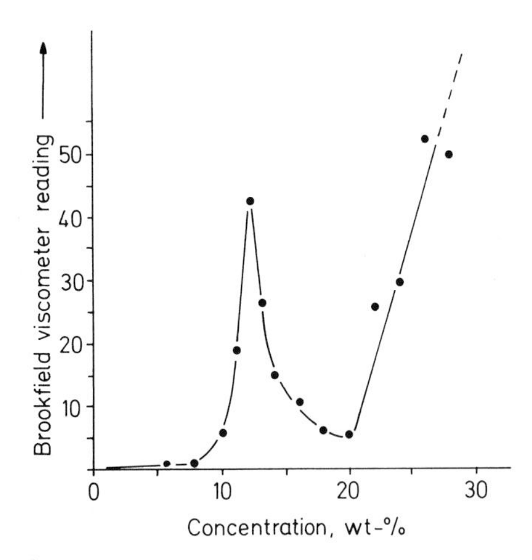

Fig. 4. Bulk viscosity vs concentration of PPD-T/H_2SO_4 solution

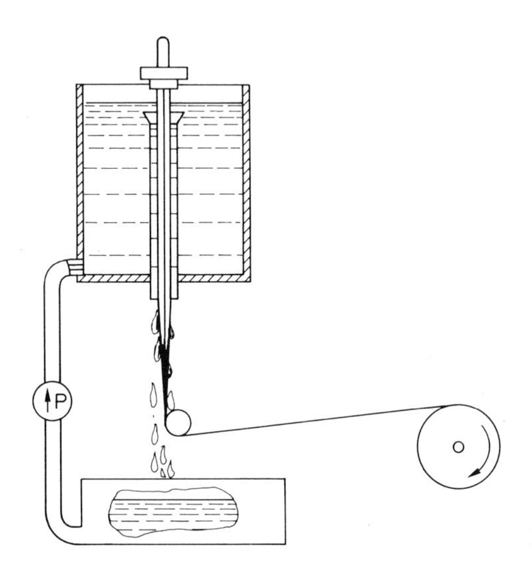

Fig. 5. Dry jet wet spinning process

3.3.1 Comparison of Different Aramid Fibers

Properties of various commercial aramids are summarized in Table 4. Water sensitivity or moisture regain of aramids varies from 1.1 to 7 weight %. Kitamaru and coworkers [24] examined water sorption of Kevlar 29 and 49 by

Table 4. Commercial aramids[a]

	KEVLAR (23)				TWARON[b]		TECHNORA[c]
	29	129	49	149	Regular	HM	
Density (kg/m^3)	1440	1440	1440	1470	1440	1450	1390
Tensile stength (GPa)	2.9	—	2.9	2.4	2.8	2.8	3.1
Tensile modulus (GPa)	62	—	114	146	80	115	81
Elongation (%)	4.0	3.3	2.8	2.0	3.3	2.0	4.4
Moisture regain (25 °C, 55% R.H.)	4.8	4.3	4.3	1.1	7	—	2.0
(20 °C, 65% R.H.)						3.5	—

[a] All 12 μm diameter
[b] AKZO brochure
[c] Teijin

analyzing polymer morphology through cross polarization/magic angle spinning (CP/MAS) ^{13}C NMR. Their studies suggest that Kevlar 49, which is higher in crystallinity than Kevlar 29 but has similar water sorption, may not be entirely crystalline but may include some non-crystalline-like fraction into which water can be adsorbed.

High toughness and high modulus aramids are available from Dupont and Akzo. Recently improved aramids, Kevlar 129 and Kevlar 149, were introduced by Dupont. Kevlar 149 is a slightly higher density (1470 kg/m^3) fiber with an attractive lower moisture regain value of 1.1 weight %. The aramid, Technora, is about 5% lower in density (1390) with higher tensile strength and intermediate modulus as compared to high modulus Kevlar 49 and Twaron.

The favorable high specific strength and specific modulus of aramids (Fig. 1) as well as high temperature resistance have been prime factors in the utilization of aramids in many high performance composites which were previously based on glass fibers. Composites requiring favorable strength to weight and stiffness to weight characteristics resulted in selecting aramids for filament wound applications, particularly motor cases in missile systems. The specific strength of a fiber compares favorably with the burst strength of a cylindrical component (PV/W where the product of burst pressure and volume is divided by its weight). However, translation of aramid tensile strength from fiber to cylindrical component was lower than glass or graphite. By developing a special fiber surface modification for the aramid, better translation of strength occurred due to optimum adhesion between fiber and matrix in the filament-wound cylinder (Fig. 6) [25].

The interfacial strength between Kevlar fiber and a matrix resin depends upon the interfacial adhesion that can be developed on the Kevlar fiber surface by the matrix resin. Through the use of grazing angle X-ray Photoelectron Spectroscopy (XPS) also known as ESCA [26], surface analysis of Kevlar suggests that nitrogen is virtually nonexistent while oxygen is higher than expected (Kevlar ratio is 14 carbons: 2 oxygens: 2 nitrogen). The carbon peak is a doublet and indicates two types of carbon and inconsistent with Kevlar

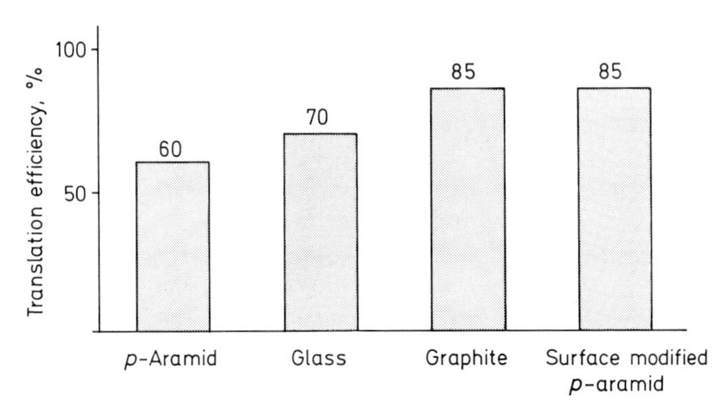

Fig. 6. Efficiency of strength translation in filament-wound cylinders

structure. Yet when Kevlar is scanned perpendicular to the analyzer axis, only a single carbon peak is observed as well as an expected Nitrogen peak. Depending on the scanning angle, different surface effects are noted for Kevlar. These anomalous surface effects may be related to difficulties that persist in improving fiber-resin adhesion [21]. Low accessibility of the NH groups of either Kevlar 29 or Kevlar 49 by ATR FTIR [27] or photoacoustic FTIR spectroscopy [28] have been reported.

Mediocre adhesion between Kevlar fiber and the matrix is highly desirable and beneficial when Kevlar composites are fabricated into armor or the U.S. Army ballistic helmet. The impedance of the projectile depends mainly on Kevlar fiber strength [29]. Kevlar can only retard the projectile threat if a rapid composite delamination facilitated by marginal adhesion occurs on impact.

Another equally important physical characteristic attributable to aramid is its toughness and its applicability to composite impact damage tolerance. In falling weight impact tests, aramid composite absorbs approximately twice as much energy without damage as carbon fiber composite [30]. A significant amount of aramid is being utilized in commercial aircraft in non-load bearing structures or in aramid/carbon fiber hybrids where higher impact damage tolerance is required.

3.4 Carbon Fibers

Introduction

Since their disclosure in the 1960s, high performance or structural carbon (graphite) fibers have become the predominant reinforcement for advanced composite systems and have enabled unprecedented advancements in the aircraft, missile, space hardware, and sporting goods industries. Watt and Perov [31] deal with the preparation, structure, and properties of strong fibers including carbon or graphite fibers. Carbon fibers based on three precursors [rayon, poly(acrylonitrile) or PAN, and mesophase pitch] dominate the industry based on their unique mechanical and physical properties. A directory, currently in the third edition, [32] has been compiled and provides a comprehensive listing of fiber suppliers by product form. Typical mechanical and physical properties are also presented in this directory. A very interesting historical account of the development of carbon fibers was presented by Bacon [33] during an ACS meeting in 1986. Fitzer's book on Carbon Fibers [34] summarized worldwide activities of CF and Composites.

The properties of the various types of carbon fibers will be presented along with important structural characteristics. Carbon fiber surface chemistry will not be discussed in detail since much of this information is proprietary to specific fiber manufacturers.

There are many carbon fiber manufacturers worldwide. Rather than providing a detailed listing of these manufacturers and the range of CF products, the product line of Amoco Performance Products, Inc., a fully integrated supplier of carbon fiber based on rayon, PAN and mesophase pitch, is presented.

3.4.1 Rayon Precursor

Rayon is thermally treated under special conditions into low modulus carbon fibers (≤ 50 GPa).

The primary application for carbon and graphite fibers based on rayon precursor is in the rocket and missile industry for rocket nozzle and other ablative applications. Reinforcements for ablative composites must provide structural integrity at the lowest weight, and maintain high reliability while keeping thermal erosion and char depth to a minimum. The critical attributes to achieve performance in these applications include: low thermal conductivity, low density, high physical thermal stability and carbon content, tensile strength and modulus sufficient for structural integrity, low sodium content which is beneficial to promoting oxidative stability and low moisture adsorptivity.

Rayon based carbon and graphite fibers are typically used in fabric form to produce phenolic matrix based and/or carbon/carbon composites for ablative

Table 5. Typical properties of rayon-based Thornel carbon and graphite fibers and fabrics

Material precursor product weave	Carbon rayon			Graphite rayon	
	VCK 5HS	VCL 8HS	VCX-13 5HS	WCA plain	
Total fabric properties					
Width, m		1.09–1.14	1.07–1.14	1.09–1.14	1.07–1.15
Weight, g/m^2		254.3	271.2	254.3	244.1
Gage (thickness)-mm		0.457	0.508	0.457	0.559
Count, Yarns/mm	Warp	1.57	2.09	1.57	1.14
	Fill	1.42	2.05	1.42	0.83
Filaments/Yarn bundle		980	720	980	1470
Carbon assay, %		99 +	97.0	99.5 +	99.9
Ash content, %		0.30	0.40	0.08	0.01
Tensile, kg/m	Warp	625.0	803.6	625.0	892.9
	Fill	535.7	803.6	535.7	357.2
Electrical resistivity (ohms/square)	Warp	0.45	0.60	0.45	0.38
	Fill	0.46	0.56	0.46	0.53
Density, kg/m^3 (ASTM D-3800)		1500	1520	1470	1440
Typical yarn properties					
Water adsorption capacity %		2	13	1	0.1
Strength, GPa		0.69	0.69	0.69	0.69
Modulus, GPa		41.4	41.4	41.4	41.4
Electrical resistivity (u-ohm-m)		38	59	36	36
Thermal conductivity (W/m K)		4.0	3.7	4.1	4.1

and nozzle applications. Typical properties are shown in Table 5 for the Thornel [35] carbon and graphite fibers and fabrics. PAN and pitch based carbon and graphite products suitable for use in ablative applications are also listed in the literature [35]. Towne [36–38] describes some recent work with carbon and graphite fabrics based on rayon that deal with rocket nozzle uses.

3.4.2 Pitch Precursor

Carbon fibers based on petroleum pitch have a unique property profile which includes: very high axial modulus (up to 896 GPa); strongly negative values of axial thermal expansion coefficient; high axial thermal (and electrical) conductivity, and adequate tensile properties. These characteristics make composites based on high modulus pitch fibers ideally suited for space hardware and avionic applications. The resulting structure possesses optimal weight, high

Table 6. Typical mechanical and thermal properties of ultra high modulus pitch fibers

Property[a]/Fiber	P-75	P-100	P-120
Moduli			
E_a, (GPa)	517	690	827
E_t, (GPa)	9	6.9	6.9
G_a, (GPa)	13	21	21
ν_a	0.23	0.26	0.3
ν_t	0.74	0.74	0.85
Strength			
σ^{tu}, (MPa)[b]	1900	2240	2240
σ^{cu}, (MPa)[c]	690	520	410
Thermal expansion			
α_a, 10^{-6}/K	− 1.46	− 1.48	− 1.50
α_t, 10^{-6}/K	12.5	12.0	12.0
Thermal conductivity			
K_a, (W/mK)	190	530	610
K_t, (W/mK)	2.4	2.4	2.4
Specific heat			
C, (kJ/kgK)	1.0	1.0	1.0
Density			
ρ, (kg/m³)	2160	2160	2190

[a] subscript a = axial; subscript t = transverse; superscript t = tensile; superscript c = compressive; superscript u = ultimate
[b] representative data from strand and/or laminate tests
[c] estimated

stiffness, high thermal conductivity, dimensional stability and adequate strength. The unidirectional compressive strength of composites based on high modulus pitch fiber is considered moderate due to the failure mechanisms inherent to the fiber. Hence, pitch fiber based composites are not ideally suited for structure driven by compressive strength. Recently Hahn [39] compared the compressive strength and failure modes of pitch fibers and PAN based CF. Lower compressive strength of pitch fibers versus PAN was attributed to microstructural differences of the fibers with pitch fibers being susceptible to shear. Recent advances in high modulus pitch fibers are documented by Schulz [40]. Pitch fiber developments are reviewed in [41–47] along with thermo-mechanical characterization of the pitches used to produce carbon fiber [48] and complimentary rheological data [49]. Studies of thermal expansion behavior of pitch fiber based composites is presented by Bacon and coworkers [50].

Pitch based carbon fibers covering a broad range of properties are designated as P-25, P-55, P-75, P-100, and P-120. They represented uniaxial fiber moduli from 172 to 827 GPa [51].

The lowest modulus pitch fiber P-25 is widely used in carbon/carbon brake applications. P-55 fiber which has a 380 GPa modulus is used in structures requiring a balance in stiffness, strength, and physical properties.

Table 6 presents typical mechanical and thermal properties of ultra high modulus pitch based fibers. Specifically, the modulus ranges from 517 to 827 GPa. Properties are listed for both the axial and transverse directions, in recognition of the anisotropy inherent to the fiber. Particularly noteworthy are the high axial modulus, high axial thermal conductivity, and the negative axial thermal expansion. It is these properties which afford stiff, conductive, dimensionally stable composite structures.

3.4.3 PAN Precursor

Earlier it was mentioned that carbon fiber is made from one of three precursor materials: rayon, mesophase pitch, and polyacrylonitrile (PAN). Rayon based carbon fibers are used chiefly in ablative and carbon/carbon applications. Pitch precursor is used for very high modulus carbon fiber. PAN is the precursor of choice for very high strength carbon fiber.

The manufacture of PAN based CF is a multi-step process commencing with the polymerization of acrylonitrile and selective comonomers to PAN, followed by heat treating (crosslinking) the PAN fiber and ultimately carbonizing it to carbon fiber. A comprehensive characterization of the chemical structure of PAN based CF from polyacrylonitrile was recently conducted by Morita [52]. Various techniques such as SEM, TEM, electron diffraction, and electron energy loss spectroscopy were employed for structure while CF surface and interface of CF/matrix resin was studied by surface and microanalytical methods, primarily XPS and complemented by Raman microprobe, FTIR, solid state NMR and SIMS. Newer advances related to PAN carbon fibers with superior oxidation resistance were reported by Kowalski [53].

The process to manufacture PAN carbon fibers is as follows. The first step in the process is to polymerize the precursor material which starts with the free radical polymerization of acrylonitrile (AN) and suitable comonomers to form PAN. Polymerization is followed by spinning the precursor polymer into a multifilament strand, usually containing between 1000 and 12000 individual filaments. PAN is typically dry or wet spun in solution. Single or multiple drawing stages are used to attenuate the fiber and improve the orientation of the polymer chains along the fiber axis. The drawing and consolidation of PAN-based fibers requires multiple stages and precise control of operating parameters. The resultant white fiber precursor visually resembles fiberglass roving. The fibers are next thermally stabilized via an oxidation process which employs temperatures approaching 400 °C. This involves crosslinking of the polymer chains to make the fiber intractable so that it will not melt when subjected to increasingly higher temperatures. The PAN-based fiber changes from white to black during stabilization. The next step of the process is carbonization which takes place in an inert atmosphere at temperatures between 1000 °C and 2000 °C. To achieve very high modulus, a graphitization step at 2000 °C to 3000 °C is used. The carbon or graphite fibers have achieved their final strength, modulus, and density at this stage of the process. Some understanding of the heat treatment of PAN fibers is provided by recent studies of Guigon and Oberlin [54]. They describe the heat treatment of HTS PAN carbon fiber by examining microtexture, mechanical properties of single filaments and elemental analyses. Their studies suggest the atomic ratio of nitrogen to carbon is related to microtexture which further is related to structure and mechanical properties. Tensile strength is higher with higher values of N/C at the highest carbonization temperature. Heat treating above 2000 °C reduces N/C and yields HM fibers with a high strength.

Balancing modulus, tensile and compressive strength is becoming more important than developing CF with higher tensile strength. Better knowledge and understanding of the chemistry and process for the manufacture of CF allows producers to tailor fibers to almost any structural requirement. Carbon fibers are being created with fewer defects, with higher compressive strength, and modulus/strength combinations for composites meeting a broad range of structural requirements. Studies dealing with fiber morphology and defect populations are ongoing activities. Recently a model [55] has been developed to establish defect populations quantitatively. Experimental tensile data on single filaments compared with resin impregnated fiber tow tensile data provide a way to predict and control defects.

To aid bonding of the fibers to composite matrix materials, a surface treatment is performed that removes a weak surface layer from the fiber and functionalizes the fiber surface to enhance chemical bonding with the matrix. The final steps in the process are the application of a sizing to aid handleability and winding of the fiber on spools.

There are three major structural parameters of a carbon fiber: orientation, crystallinity, and defect content (Fig. 7).

Orientation is an average measurement of how well the graphite layer planes are aligned with the axis of the fiber. Crystallinity is a combined measurement of both crystalline perfection and crystallite size. The degree of perfection is related to how close the parallel spacing of the graphite layer planes approach the theoretical value of 3.35 Å. The crystallite size is the average stack height of parallel graphite layer planes. Both orientation and crystallinity are measured by X-ray diffraction. Defect content of carbon fibers is easily seen by examining the fracture surfaces of carefully broken single filaments. Defects can be internal (voids, inclusions) or external (pits, gouges, extraneous material).

Degree of orientation in carbon fibers has a major effect on the final carbon fiber properties. In general, increasing the degree of orientation improves the longitudinal properties and decreases the transverse properties. Specifically, longitudinal tensile strength, tensile modulus, thermal conductivity, electrical conductivity and the negative CTE are increased. In the transverse direction, both tensile strength and tensile modulus decrease.

As the degree of crystallinity is increased many carbon fiber properties are affected. The following properties increase: thermal conductivity, electrical conductivity, the longitudinal negative CTE, and oxidation resistance. The following properties decrease: longitudinal tensile strength, compressive strength and shear modulus, as well as the transverse strength and modulus.

The removal of defects from carbon fibers leads only to beneficial effects. As defects are removed, tensile strength, thermal conductivity, electrical conductivity, and oxidation resistance are all increased.

The carbon fiber structural parameters are controlled by a number of processing steps. The fiber orientation is controlled by the fiber spinning, fiber

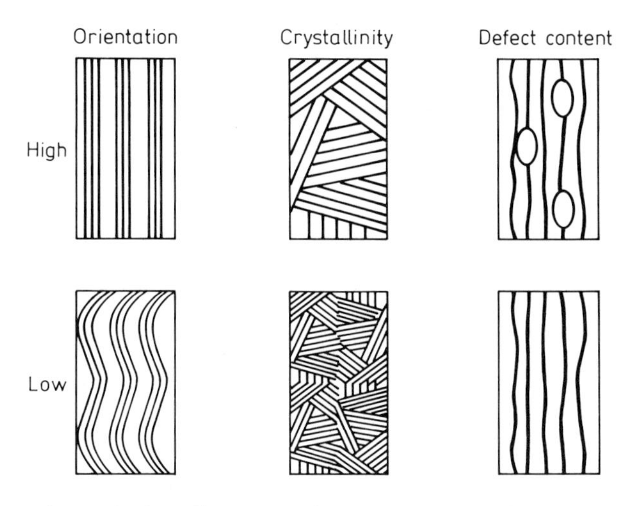

Fig. 7. Carbon fiber properties are determined by their structure

Table 7. Typical properties Thornel PAN-based carbon fibers

	T-300	T650-35	T650-42
Axial[a] tensile strength, (GPa)	3.65	4.55	5.03
Axial tensile modulus, (GPa)	231	248	290
Density, (kg/m^3)	1760	1770	1780
Filament Diameter, (μ)	7	6.8	5.1
Elongation at break, (%)	1.4	1.75	1.7
Carbon assay, (%)	92	94	94
Surface area, (m^2/g)	0.45	0.5	0.5
Axial thermal conductivity, (W/mK)	8.5	14	15
Electrical resistivity, (μ ohm-m)	18	14.9	14.2
Axial CTE at 21 °C, (ppm/°C)	– 0.6	– 0.6	– 0.75

[a] Axial and longitudinal are used interchangeably throughout.

drawing , and the heat treatment. The crystallinity is governed by the precursor chemistry and the heat treatment. The defect content is controlled by the precursor polymer purity, by careful handling, and protection from contaminants throughout the process.

In order to produce the desired carbon fiber structure and, thereby, control the final properties, it is important to have control over all of the process elements from manufacture of polymer through spinning, drawing, and final heat treatments. A fully integrated operation from AN monomer to CF is performed by Amoco Performance Products, Inc. in Greenville, SC.

Techniques for measurement of carbon fiber properties include a combination of measurements of individual fibers or fiber bundles. It can also be a measurement of composite properties with an indirect determination of the fiber property [46]. Techniques and data are presented by several authors [56–58]. Eckstein describes thermal oxidative measurements as well as detailed nonlinear stress strain behavior of carbon fibers and composites [59].

A listing of typical PAN based carbon fiber properties are contained in Table 7 [60].

3.5 S-Glass

Since its introduction in 1939, glass fiber has gained wide acceptance in general purpose composite materials. General purpose glass known as "E Glass" contains a calcium aluminoborosilicate composition while S Glass is a magnesium aluminosilicate composition and has higher tensile, modulus and compressive strength with a corresponding lower density as compared to E glass [61]. S-Glass and S-2 Glass are compositionally similar but possess different surface coatings [62].

3.5.1 Comparison of S-Glass and E-Glass

A comparison of the physical properties of E- and S-Glass is shown in Table 8. Comparing specific strength and specific modulus of S Glass, S-Glass has a specific strength value of 18×10^6 cm and within the guidelines of a high performance fiber but is lacking in specific modulus with a value of about 4×10^8 cm and below the minimum value of 6.5×10^8 arbitrarily set for high performance fibers. Nevertheless several high performance applications have been developed with S-2 Glass as the preferred fiber. These include ballistic armor for U.S. Navy vessels and U.S. Army vehicles, and for use in aircraft cargo liners. Early filament wound rocket motor cases and pressure vessels were composed of glass but were later displaced by stiffer and stronger fibers.

3.6 Other Fibers

Newer high strength fibers are based on rigid rod polymers and liquid crystalline polymers. The rigid rod polymers are known as PBO and PBT. These materials possess excellent thermal stability, stiffness and strength.

Synthesis of both is based on the reaction of terephthalic acid with 2,4 diamino resorcinol for PBO or 2,5 diamino-1,4 benzene dithiol to PBT.

Spinning of these fibers is conducted from acid solutions such as polyphosphoric acid or methanesulfonic acid via a dry jet-wet method.

The preliminary high strength and modulus values for these rigid rod fibers coupled to their low density results in high specific strength and modulus values. Like other organic fibers such as aramids or UHMWPE they are deficient in compressive strength. This particular physical characteristic limits their utility for any load bearing composite structures.

The morphology and property differences between fiber surface and fiber interior are frequently referred to as a "skin-core" effect and are commonly observed for PBT, PBO and aramid type fibers.

Possible methods to overcome several of these deficiencies (poor compressive strength and skin-core effect) may require better fiber spinning techniques for defect-free fibers, novel surface treatment or the formation of non-circular fiber cross section. The latter technique has been beneficial in improving the performance of selected carbon fibers.

Table 8. Comparison of properties of E- and S-glass

	E-glass	S-glass
Density (kg/m^3)	2580	2490
Tensile strength (GPa)	3.45	4.6
Tensile modulus (GPa)	72.5	85.0
Elongation (%)	4.9	5.7

3.7 Summary of High Performance Fibers

A variety of high performance fibers with densities as low as 950–2500 kg/m^3 with impressive specific strengths and specific moduli are described. PAN based carbon fibers are the pre-eminent fiber of choice as reinforcing agent in ACM applications requiring high strength/stiffness and compressive strength characteristics. Organic fibers such as UHMWPE and Aramid provide toughness and reduced density and are important fiber components in selective ballistic applications such as helmets and vests. S-Glass with its high specific strength is the preferred fiber component for military armor for U.S. naval vessels and U.S. Army vehicles. Newly developing rigid rod organic polymer fibers exhibit excellent thermal stability, stiffness, and strength but are lacking in compressive strength.

3.8 References

1. H. S. Matsuda, Chem. Tech. (May) 310 (1988).
2. S. Kumar, SAMPE Quart. 20(2), 3 (1989).
3. W. W. Adams, Poly. Preprints, 26(2), 306 (1985).
4. J. J. Smith, H. Jiang, R. K. Eby and W. W. Adams, Poly. Commun. 28, 14 (1987).
5. Y. Tremonia and P. Smith, Chapter 11, in "High Modulus Polymers," A. E. Zachariades and R. S. Porter, editors, Marcel Dekker, NY 1988.
6. T. Y. Tam, M. B. Boone and G. C. Weedon, Poly. Eng. & Sci. 28(13), 871 (1988).
7. A. J. Pennings et al., in "Interrelations Between Processing, Structure, and Properties of Polymeric Materials," J. C. Seferis and P. S. Theocaris, editors, p. 285 Elsevier, Oxford 1984.
8. Polymer News, 14(3), 83 (1989).
9. J. H. Southern and R. S. Porter, J. Appl. Poly. Sci., 14, 2305 (1970).
10. G. Capaccio and I. M. Ward, Poly. Eng. Sci., 15, 219 (1975).
11. W. Hoogsteen, G. ten Brinke and A. J. Pennings, Colloid Polym. Sci., 266, 1003 (1988).
12. J. P. Cohen-Addad, G. Feio and A. Peguy, Poly. Commun. 28, 252 (1987).
13. M. Matusuo et al., Poly. J. 18(10), 759 (1986).
14. R. J. Samuels, J. Macromol. Sci.-Phys. B4 701 (1970).
15. Y. D. Kwon, S. Kavesh and D. C. Prevorsek, U.S. 4,713,290 (12/15/87) Allied.
16. G. A. Harpell, I. Palley, S. Kavesh and D. C. Prevorsek, U.S. 4,737,401 (4/12/88) Allied.
17. D. J. Dijkstra, W. Hoogsteen and A. J. Penning, Polymer 30, 866 (1989).
18. G. B. Kaufman, Chem. Tech. 725 Dec. (1988).
19. M. Jaffe and R. S. Jones, "Handbook of Fiber Science and Technology," Vol III, High Technology Fibers, Part A, Marcel Dekker, NY. 1985.
20. L. Vollbracht, Chapter 22 in "Comprehensive Polymer Science," Vol 5, Sir G. Allen, J. C. Bevington, editors, Pergmon, Oxford 1989.
21. H. H. Yang, "Aramid Fibers," Chapter 6, in "Composite Materials," Series 2, Fiber Reinforcements for Composite Materials, R. Bunsell, editor, Elsevier, Amsterdam 1988.
22. H. Blades, 3,869,429 (1975) Dupont.

23. P. G. Riewald, A. K. Dhingra and T. S. Chern, 6th Int'l Conf. on Comp. Materials and 2nd European Conf. on Com. Mat. Vol 5, Editors, F. L. Matthews. N. C. R. Buskell, J. M. Hodgkinson and J. Morton, Elsevier, London 1987, p. 362.
24. M. Fukuda, H. Kawai, F. Harii and R. Kitamaru, Poly. Commum. 29, 97 (1988).
25. D. Tanner, J. A. Fitzgerald, P. G. Riewald and W. F. Knoff, Handbook of Fibers. Science and Technology, Vol. III, High Technology Fibers, Part B, Marcel Dekker, NY 1989.
26. C. D. Batich, P. H. Holloway and M. A. Kosinski, CHEMTECH, Aug. p. 494, 1986.
27. F. Druschle, H. W. Siesler, G. Spilgies and H. Tengler. Poly. Sci. Eng., 17(2), 93 (1977).
28. E. G. Chatzi, M. W. Urban, H. Ishida and J. L. Koenig, Polymer 27, 1850 (1986).
29. C. E. Morrison and W. H. Bowyer, "Proceedings of the 3rd National Conference on Composite Materials," Paris, France August 26–9, 1980, pp. 233–245.
30. M. W. Wardle and E. W. Tokarsky, Comp. Technol. Rev. ASTM, Philadelphia, Spring 1983.
31. W. Watt and B. V. Perov, Strong Fibers, Handbook of Composites, Vol 1, Elsevier Science, 1985.
32. D. Pamington, (editor) Carbon & High Performance Fibers Directory – 3rd edition, PAMMAC Directories.
33. R. Bacon and C. T. Moses, High Performance Polymers: Their Origin and Development, Proceedings of a Symp. held at 191st ACS meeting, New York, NY, April 15–18, 1986, pp. 341–353.
34. E. Fitzer, editor, "Carbon Fibers and their Composites," Springer, Berlin Heidelberg New York (1984).
35. Amoco Performance Products, Inc., Product Literature F7238, 1989.
36. M. K. Towne, Proceedings of the 7th JANNAF Rocket Nozzel Technology Subcommittee Meeting, Naval Postgraduate School, Monterey, CA, Nov. 12–14, 1985. Available from CPIA, Johns Hopkins University, Applied Physics Laboratory, Johns Hopkins Road, Laurel, Maryland 20707.
37. M. K. Towne, Proceedings of the 7th JANNAF Rocket Nozzel Technology Subcommittee Meeting, Naval Postgraduate School, Monterey, CA, Nov. 12–14, 1985. Available from CPIA, Johns Hopkins University, Applied Physics Laboratory, Johns Hopkins Road, Laurel, Maryland 20707.
38. M. K. Towne, "Quality and Product Improvements on Rayon-Based Carbon and Graphite Fabrics," Proceedings of the 7th JANNAF Rocket Nozzel Technology Subcommittee Meeting, Naval Postgraduate School, Monterey, CA, Oct. 19–22, 1982.
39. J. M. Prandy and H. T. Hahn, 35 International SAMPE, April 1990, p. 1157.
40. D. A. Schulz, SAMPE Journal, 23:27–31, 109 (March/April, 1987).
41. R. Bacon, Metal and Ceramic Matrix Composite Processing Conference Proceedings, Battelle, Columbus Laboratories, Nov. 13–15, 1984, Vol. 2, pp. 23–28.
42. L. S. Singer, Extended Abstracts of the International Symp. on Carbon: New Processing and New Applications, Toyohashi, Japan, November, 1982, Paper 4A01, pp. 400–403.
43. L. S. Singer, Fuel, 60:839–847, Sept., 1981.
44. R. Bacon, Phil. Trans. Royal Soc. (London) A, 294:437–442. (1979).
45. L. S. Singer, A. Ciferri and I. M. Ward, Applied Science, 1979, pp. 251–277.
46. A. A. Bright and L. S. Singer, Carbon, 17:59–69 (No. 1, 1979).
47. S. Chwastiak, J. B. Barr and R. Didchenko, Carbon, 17:49–53, 1979.
48. J. B. Barr, I. C. Lewis, Thermochim. Acta., 52:297–304, Jan 16, 1982.
49. F. F. Nazem, Pitch, Fuel, 80:851–858.
50. R. Bacon, B. H. Eckstein, I. M. Kowalski, D. A. Schulz, S. L. Strong, M. K. Towne and G. Wagoner, Proceedings of the First European Conference on Composite Materials and Exhibition European Association for Composite Materials, Bordeaux, France, Sept., 25–27, 1985, pp. 152–157.

51. Amoco Performance Products, Inc., Product Literature, F-5869 Rev. 6, F-7144, F-7145, F-7146, F-7149, 1988.
52. K. Morita et al., Pure and Applied Chem. 58(3), 455 (1986).
53. I. M. Kowalski, Anaheim, CA, April 6–9, 1987, pp. 953–963.
54. M. Guigon and A. Oberlin, Composites Science & Technology 27(1), (1986).
55. S. B. Smith, 34th International SAMPE Conf. May 1989, p. 1621.
56. G. Wagoner and R. Bacon, Proceedings of the 19th Biennial Carbon Conference, Penn State University, June 26–30, 1989, pp. 296–297.
57. G. Wagoner, R. E. Smith and R. Bacon, Proceedings of the 18th Biennial Conference on Carbon, Worcester, MA, July 19–24, 1987, pp. 415–416.
58. I. M. Kowalski, 31st International SAMPE Symp. and Exhibition, Las Vegas, Nevada, April 7–10, 1986, pp. 303–314.
59. J. B. Barr and B. H. Eckstein, Proceedings of the 18th Biennial Conference on Carbon, Worcester, MA, July 19–24, 1987, pp. 9–10.
60. Amoco Performance Products, Inc., Product Literature, F-7147, F-7178, F-7150, 1988.
61. D. Miller, "Glass Fibers" in "Composites," Vol 1, ASM International, 1987.
62. R. W. Fulmer, SAMPE, May 1980.

4 Analysis/Testing

4.1 Introduction

The salient features of Advanced Composite Materials such as low composite weight coupled with high strength and excellent fatigue resistance will only result if the fabricated multi-ply laminate is assembled with the utmost diligence. The anisotropic nature of composites suggests that they are more process sensitive than isotropic materials. Subtle defects or flaws may occur as a result of material fabrication or through "in-service" use. These defects may upset the unique balance of fiber, matrix, and interface that provides the composite with high performance characteristics. Quality control during various phases of manufacture and assembly is critical. Analyses of different stages from raw materials to finished part are essential to ensure the integrity of the composite part. During the analysis of raw materials to the intermediate stage (prepregs), sacrificial analytical methods are commonly used whereas the final component or composite part is analyzed via a non-invasive or non-destructive test (NDT), a methodology that allows analyses to be conducted without damaging the functional capability of the object under test.

4.2 Fiber

Fibers are well characterized by fiber manufacturers. Sufficient information is provided by the manufacturers since most physical and mechanical properties are determined during the early stages of fiber commercial development. Physical properties such as fiber density, diameter, tenacity, modulus, stress/strain, moisture absorption, solvent and environmental resistance and maximum temperature use are typical data that provide sufficient fiber guidelines in selecting reinforcing agents for ACM.

4.2.1 Surface/Structural Methods

Additional information is necessary for fully describing fibers. Fiber surface characteristics which relate to strength, modulus via orientation can be examined by XPS (Table 1). X-Ray Photoelectron Spectroscopy (XPS), also known as ESCA, has been used to investigate fiber finishes and the depth of penetration of finishes into fiber interior [1].

Table 1. Surface/Structural methods

	SEM/ TEM	XPS	ISS/ SIMS	STM	DRIFT ATR-FTIR	SAXS	SANS	NMR	RS	PAS
Elemental Composition (M, F, C)		*	*					*		
Surface Analysis of Fibers/Polymers (M, F, C)		*			*			*	*	
Interfacial Bonding Fiber to Matrix (C)	*			*						*
Crack Growth/Fracture Initiation (C)	*									
Surface Morphology/Chemistry (M, F, C)	*	*	*	*				*	*	
Polymer Impact Modifiers (M, C)						*		*		
Sizing/Coupling Agents (F)					*					
Blend Phase Separation (M)						*				*
IPN Phase Analysis (M, C)							*			*
Polymer Composition (M, F, C)			*				*	*		*
Fiber Deformation (F, C)									*	

SEM/TEM Scanning Electron/Transmission Electron Microscopy
XPS X-Ray Photoelectron Spectroscopy (ESCA)
ISS/SIMS Ion Scattering/Secondary Ion Mass Spectroscopy
STM Scanning Tunnelling Microscopy
DRIFT Diffuse Reflectance IR Fourier Transform Spectroscopy
ATR-FTIR Attenuated Total Reflectance Fourier Transform IR Spectroscopy
SAXS Small Angle X-Ray Scattering
SANS Small Angle Neutron Scattering
NMR Nuclear Magnetic Resonance
RS Raman Spectroscopy
PAS Positron Annihilation Spectroscopy [21]
M Matrix
F Fiber
C Composite

Attenuated Total Reflectance FTIR (ATR-FTIR), and Diffuse Reflectance IR Fourier Transform Spectroscopy (DRIFT) have provided information as it relates to orientation effects of Kevlar 49 surface [2] and surface hydrolysis [3]. Fiber deformation of aramids [4], carbon fibers and composites [5] are related to applied strain and can be monitored by Raman Spectroscopy. Water absorption in Kevlar 49 has been studied by NMR (CP/MAS ^{13}C) and suggests that a small portion of a non-crystalline like fraction of Kevlar may be responsible for water sorption [6].

4.3 Resin/Prepreg

The fiber as a uni- or multidirectional tape or woven fabric is preimpregnated with resin to yield an intermediate product known as "prepreg". High modulus matrix resins can be either thermoset or thermoplastic systems.

4.3.1 Thermoset

Resin characterization is required particularly if the desired resin is a thermoset. Since these thermoset resins are low MW oligomers, they undergo some MW advancement during the impregnation operation.

4.3.1.1 Chromatography and Thermal Analysis Method

Microstructure characterization of the thermoset resin (epoxy, phenolic, imide, BMI or cyanate) can be conducted by several types of chromatography (Table 2), thermal analyses (Table 3), as well as structural/surface methods (Table 1).

Table 2. Chromatography

	HPLC	GPC	TLC	SEC
Monomer Purity	*		*	
Oligomer Purity/Composition	*	*	*	
Polymer Analysis		*		*
Molecular Weight Distribution		*		*

HPLC High Performance Liquid Chromatography
GPC Gel Permeation Chromatography
TLC Thin Layer Chromatography
SEC Size Exclusion Chromatography [22]

Table 3. Thermal analysis

	DTA	DSC	TGA	TMA	DMA	DEA	FTIR
Glass Transition Temp. (T_g)	*	*		*	*	*	
Crystallization Temp. (T_m)	*	*				*	
Crystallinity		*				*	
Coefficient of Expansion				*			
Chemical Stability		*	*			*	*
Thermal Stability			*				*
Mechanical Stability					*		
Modulus of Elasticity/Loss Modulus					*	*	
Viscosity					*	*	
Cure Kinetics/Degree of Cure		*				*	
Solids Integrity/Voids						*	

DTA Differential Thermal Analysis
DSC Differential Scanning Calorimetry
TGA Thermogravimetric Analysis
TMA Thermomechanical Analysis
DMA Dynamic Mechanical Analysis
DEA Dielectric Analysis
FTIR Fourier Transform IR Spectroscopy (Evolved Gas)

Oligomer composition, purity and molecular weight characteristics can be determined by several chromatographic methods (HPLC, GPC or SEC). Thermal analyses methods can monitor oligomer viscosity and degree of cure (DMA or DEA), determine T_g (DSC, DMA, DEA), mechanical stability and properties (DMA). Time–Temperature–Transformation (TTT) cure diagrams can be developed and have been reported for epoxies [7] and polyamic acid-imide systems [8]. Solution or solid state NMR analyses have broad applicability in identifying oligomers, reaction intermediates and cured resin structures by examining ^{1}H, ^{13}C, ^{15}N, ^{19}F, ^{29}Si, and ^{31}P nuclei [9].

Rapid "in line" monitoring of thermoset cure can be accomplished by FTIR [10].

4.3.2 Thermoplastic

Thermoplastic resins are conveniently examined by GPC for MWD, NMR for structure, polymer morphology and/or sequence units, DMA for mechanical properties, DSC or X-Ray Diffraction for level of crystallinity, SEM/TEM for adhesion between matrix and fiber, SAXS and SANS for polymer structure.

4.3.3 Blends/IPN

Detailed identity of phase phenomenon of blends and IPN can be determined by SAXS, SANS.

4.4 Composite

A more detailed analyses of the composite is required and supercedes the usual determination of physical, mechanical, and environmental properties. These latter properties provide a comprehensive analysis of the composite and its potential physical performance for the intended application. The design features of the composite are expected to conform within the predicted limits of the composite properties.

Besides weight, strength and fatigue characteristics, the additional motivation for selecting composites is due to longevity in use. Longevity or durability is related to fatigue and environmental factors. Metal fatigue is well characterized whereas composite fatigue is quite different than metal fatigue. Further environmental effects related to temperature extremes such as tropical to subzero temperatures for aircraft or temperatures near 500 °C for engine surfaces concern differential expansion and moisture effects of fiber and matrix resin within the composite. The composite part is also susceptible to "in-service" damage from impact loading which is unavoidable or inevitable.

It was mentioned earlier that defects and flaws may occur as a result of composite manufacture or through "in-service" use. The determination of origin, detection, inhibition or elimination of source/effect of defects in composites continues to be a formidable challenge to material scientists. These defects which may be randomly distributed in the structure become associated with the onset of failure. The strength of the composite progressively deteriorates by the growth and accumulation of these microfailures. The growth of these microfailures can occur in many divergent paths throughout the composite and makes analysis of these defects/flaws and microfailures quite tedious and costly. It is not sufficient to solely ensure the integrity of the composite but it is also highly desirable to determine and predict service life/fatigue of the part.

Failure modes of composite materials vary and relate to composition. Metals undergo a ductile failure while carbon composites exhibit a brittle failure. Glass composite failure is less severe due to the higher ultimate elongation of glass.

4.4.1 Destructive Tests

Determination of the onset of failure is highly desirable for composites. Failure analysis (FA) methodology has been widely used for metals and has been utilized as a guide in developing composite failure analysis. Failure analysis can be either destructive or non-destructive. As the former method implies, the destructive test subjects the composite to a dynamic failure test mode to obtain data or information. Non-destructive analysis consists of a non-invasive technique to test the mechanical integrity without damaging functional utility. A listing of destructive tests is as follows:

- Impact/toughness – energy required to fracture a test specimen in a specified manner via shock loading.
- Deply Technique – describes the nature of fiber fracture in the composite interior.
- Fatigue/Damage – deterioration of mechanical properties after repeated application of stress with eventual damage or failure due to a large number of cycles.
- Fire/Smoke/Toxicity – apparent or lack of combustibility of composite components and potential emission of toxic gases and/or smoke. Combustibility and emission of toxic gases and/or smoke are special criteria that apply primarily to aircraft interior, cargo liners, vehicle and marine vessel interiors.

Researchers associated with various industrial and academic laboratories have developed testing protocol and different types of equipment for flammability testing. Gann [11] at the National Institute of Standards and Technology (NIST Center for Fire Research) and Hilado [12] have published extensively regarding polymer flammability. Fire/Smoke/Toxicity as it relates to Advanced Composite Materials, particularly for aircraft interiors, is receiving a significant amount of R & D effort from aircraft manufacturers, Federal Aviation Administration (U.S.), and interior panel fabricators. Sarkos [13] of the FAA has participated in aircraft interior composite flammability studies as a means of improving fire safety of aircraft interior materials as well as improving post-crash fire survivability of passengers. Testing of aircraft interior materials (see aircraft interior application) under post-crash fire conditions have led to the identification of fire/hazard progression in post-crash fire environment with emphasis on post-flashover conditions and factors affecting survivability. Survivability is judged by time to flashover and the extent of smoke toxicity. Full scale fire testing using a C-133 wide body aircraft has assisted in developing proposed new FAA low heat/smoke release test requirements for interior panels such as sidewalls, ceiling, stowage bins and partitions.

The Ohio State University (Columbus, Ohio) heat release apparatus developed by Professor Smith of OSU [14], exhibits the optimum correlation with model fire test results. It is known as ASTM E-906-83 and is used to determine time to ignition, smoke-release rates, and total heat release of various aircraft interior materials. Aircraft manufacturers, the FAA, and panel fabricators utilize the test/equipment.

A newer instrument, Cone Calorimeter, [Custom Scientific, Cedar Knolls, N.J.] measures the heat release ratio due to oxygen consumption rather than temperature rise as is determined by the OSU apparatus. Other features of the Cone Calorimeter include: a greater variety of specimens can be tested, analysis of emitted gas, and can be self-calibrated with a gas burner.

4.4.2 Non-Destructive Tests

Non-Destructive Test (NDT) methods overlap considerably, and in many cases, more than one test is employed to compliment each test and provide better defect analysis. The two main NDT methods are Sound/Sonics and X-Ray.

4.4.2.1. Sound/Sonics

The sound technique can be further divided into ultrasound and acoustic emission. Ultrasound and acoustic emission techniques are the most popular methods for evaluating composite defects such as voids/porosity, delamination or matrix/fiber cracks.

(a) Ultrasound is a method whereby high frequency sound waves are transmitted through the volume of a composite material and measured by examining attenuated signals [15]. The acoustic energy transmitted through the composite is absorbed, reflected, refracted or scattered by internal structural characteristics. If ultrasound travels unobstructed, the resulting image is bright. Ultrasound is attenuated by internal imperfections such as porosity, delamination, cracks or voids. After signal processing these imperfections become dark image areas that are interpreted visually. A liquid medium usually water, is used in ultrasonics. Depending on composite specimen size, either water immersion or water spray techniques are utilized in ultrasonics. Images can be displayed in 3 forms: A, B, and C scans. Ultrasonic C scan can detect void content as low as 1% and delaminations as small as 5 mm. It is used routinely to detect voids/porosity in ACM.

The ultrasonics technique has also been applied to the determination of fiber/resin ratio of prepregs and cured laminates [16]. An instrument with this capability is available from T.E.S.T., San Diego, CA, and requires about one minute to determine fiber/resin ratio for laminates and approximately 3 minutes for prepregs.

Newer types of fully automated ultrasonic instrumentation include Automatic Ultrasonic Scanning System (AUSS-V) from McDonnell Aircraft – (St. Louis, MO) – with a frequency range of 1 to 15 MHz. The system uses a water spray system that is directed to the composite location under analysis. Ultrasonic data are processed and displayed rapidly on a CRT in shades of gray or color. Hard copy printout is also available.

(b) Another sound technique that employs high frequency acoustic signals is Acoustic Emission (AE). Acoustic Emission [17] depends on a stressed composite to emit a signal that is evaluated. When the rigid composite is subjected to stress, various acoustical events occur on a microscopic level as the structure seeks to relieve the stress. The AE is mainly inaudible because it is too high in frequency or too low in amplitude. Detection is determined through the use of piezoelectronic sensors attached to the composite.

Different events can occur internally within the composite under stress and can relate to "debonding" between matrix and reinforcing agent, between lamina, or subtle matrix cracking/fiber breakage phenomena. Depending on the amount of stress that is applied to the composite sample, three different regions associated with AE are described [18]:

1. Under low stress the *first region* exhibits a linear, reversible mechanical behavior.
2. With increased load, initiation of irreversible damage (microcracking of matrix, fiber/matrix separation, slight delamination, fiber "pullout" or breakage) represents the *second region* and identifies damage without macroscopic propagation.
3. The *third stage* of composite under stress is the advanced damage phase and ultimate failure.

The potential to distinguish the second and third stages is critical to the understanding and use of composites. A composite in the second stage can continue to function with increasing loads (and growing damage) while in the third stage, continued stress leads to failure.

Acoustic Emission studies suggest each defect (matrix cracking, delamination, etc.) has its own energy level and relates to a range of AE amplitudes [19]. Identifiable defects with specific AE amplitudes include thickness variations or thin areas, inclusions such as solids or voids, chemical crazing, star cracks, delaminations, noncoupled second bond, dry spots.

The AE method is also effective in monitoring the integrity of a composite structure during autoclave cure. Wai and St. Germain [20] determined the AE temperature range in which filament wound tube was not susceptible to microcracking delamination and what factors in composite design contribute to microcracking.

The value of AE as an effective NDT technique is formidable. Composite defects can be located with data determined in real time. AE can determine the strength of matrix-fiber interface, fiber microbuckling, fiber stiffness. It provides on-site testing, where applicable, for composite structures such as pipes, pressure tanks/vessels and for critical composite parts in aerospace, automotive and structural market areas. The AE technique exhibits the level of damage and a potential estimate of remaining product lifetime. It assists in optimizing material selection and design by identifying predominant failure modes and the effects of modification.

Within the Society of Plastics Industry (SPI) and ASTM (U.S.), committees have been established for adapting AE equipment, techniques and AE test methods. It is speculated that AE will become one of the most important inspection techniques for component design and reliability.

An advanced acoustical microscopy method known as Scanning Laser Acoustic Microscope (SLAM) by Sonocon, Bensenville, IL, uses a focused laser beam to scan the sample area at the rate of 30 images/second with each image consisting of 40 000 image points. SLAM can locate, size and differentiate flaws in metal, ceramic, polymer and composite materials.

A closely related acoustical technique with optical display is a method known as Acoustography. Through the use of an acoustically sensitive liquid crystal display, acoustic energy emitted by the sample is converted into visual images without an intermediate line scan process. The fully automatic AIS-U acoustograph by Raj Technology, Inc., Morton Grove, Illinois, provides a instant visual image with exceptionally fine resolution. Flat, moderately curved or corrugated composite surfaces can be examined with visual identity of defects.

4.4.2.2 X-Ray

Like ultrasonics, X-ray radiography is an effective NDT technique that leads to detailed information of composite defects (voids, interply porosity, honeycomb adhesive defects, foreign materials, etc.). A variety of penetrating particles and rays such as neutron, gamma and X-ray are used to examine composites. X-ray is preferred because it is relatively easy to use and most medium size laboratories possess X-ray equipment.

Radiography is frequently used to monitor manufacturing defects like misalignment or improper prepreg stacking sequence, movement of layers prior to cure. It is quite effective in examining inserts and tabs on honeycomb panels. Liquid penetrating agents which enhance the image are often used with radiographic methods. These penetrators can be vivid dyes or X-ray opaque dyes such as selected halogenated compounds.

New improvements of X-ray technology consist of real-time X-ray imaging. Real time radiography has the capability of interfacing with image processing of automated inspection systems and allows inspection of parts in motion. An automated system whose image quality rivals that of X-ray film is Lockheed's HERTIS (High Energy Real Time Inspection System). The HERTIS operates at assembly line speeds with 100% inspection of critical parts such as castings, turbine blades, solid rocket motors and ordnance.

A portable system known as the Microfocus Unit from Lixi (Downers Grove, IL) can be taken to the source of the problem either on line during manufacturing or to the large, immobile composite object. The Microfocus Unit is particularly appealing for examination of aircraft honeycomb.

4.4.2.3 Interferometric Methods

Newer methods based on interferometric inspection systems are Holography, Electronic Speckle Pattern Imagery (ESPI) and Electronic Shearography. Holography detects composite defects by a split laser beam which undergoes cancellations and interference to create fringe lines on a film. Defects such as inhomogeneities or breaks under stress affect the composite. Comparison of stressed and unstressed composite holograms can identify inhomogeneities with minimal stress. An increase or decrease of sample temperature by one or two degrees or by changing pressure by a few tenths of a millibar provides sufficient stress. NDT uses of holography include inspection of facings on brake/clutch pads for unbonded areas, inspection of rocket engines to determine whether no debonds occur between solid fuel and rubber insulation resulting in uneven burning. Other critical applications such as aircraft honeycomb panels are inspected by holography.

Combining interferometry with video electronic equipment such as a television camera and an image storage system instead of film leads to a method known as Electronic Speckle Pattern Interferometry (ESPI). The "speckle" designation is derived from the speckled appearance of laser illumination when examined through a lens. The electronically stored hologram image is compared with stressed images and by arithmetic difference of stressed and unstressed images, light and dark patterns attributable to deformation are obtained. ESPI resolution is not as effective as a hologram. ESPI is more useful for vibration analysis. It is a rapid method with an image recorded in about 3 milliseconds.

Electronic Shearography measures interference between two slightly shifted object images without a reference beam. The method is not affected by vibration. Like ESPI the results are processed electronically and rapidly. Honeycomb composites in the B-2 Stealth Bomber are examined by this technique as well as the inspection of AWAC aircraft radomes for crushed cores and debonds with only 7×10^{-3} kg/cm^2 stress. Electron Shearography systems are available from Laser Technology, Inc., Norristown, PA.

There are several advantages associated with interferometry. These include: speed and the large area which can be scanned instantly; scan area size is limited by the laser's power to illuminate inspected area onto film or electronic image; and coherence of the laser beam facilitates illumination of increased area. For example by increasing laser power from 25 mW to 5 W, a greater range results: 25 mW provides < 0.1 m^2 at a distance of about 1 meter while the 5 W laser inspects 1.5 m^2 at a distance of about 3 meters. Further advantages include the absence of coupling liquids which are common for sonics or radiography and no contact with the composite part is required. Interferometry is an excellent technique for the detection of honeycomb defects.

4.4.2.4 Other Techniques

Other NDT methods used in the study of defects in composites include surface topography or edge replication, stiffness measurement for monitoring the degree of damage in composites, and thermography which is surface mapping of isothermal contour lines.

4.5 References

1. C. Q. Yang, J. F. Moulder and W. G. Fately, Poly. Reprints 28(1), 3 (1987).
2. A. M. Tiefenthaler and M. W. Urban, ibid., 28(1), 6 (1987).
3. E. G. Chatzi, S. L. Tidrick and J. L. Koenig, ibid., 28(1), 13 (1987).
4. R. J. Young et al., Poly. Commun. 28, 276 (1987).
5. R. J. Young, Brit. Poly. J., 21, 17 (1989).
6. M. Fukuda, H. Kawai, F. Horii and R. Kitamaru, Poly. Commun. 29, 97 (1988).
7. J. K. Gillham, Makromol. Chem. Macromol. Symp. 7, 67 (1987).
8. G. R. Palmese and J. K. Gillham, J. Appl. Poly. Sci., 31, 1925 (1987).
9. J. N. Shoolery, American Laboratory, p. 49–61 October 1988.
10. M. R. Coates, Modern Plastics, p. 90, May 1986.
11. R. G. Gann and M. J. Drews, "Flammability" in Encyclopedia of Poly. Sci. and Tech. 2nd Edition, Vol. 7 editors, H. F. Mark, N. M. Bikales, C. G. Overberger, G. Menges and J. I. Kroschwitz, Wiley, NY 1987, p. 154.
12. C. J. Hilado, "Flammability Handbook for Plastics," 3rd Edition, Technomic Publishing, Westport, CT 1982.
13. C. P. Sarkos, "Proceedings of the Aircraft Interior Materials/Fire Performance," April 4–5, 1989, Wichita State University, Kansas.
14. E. Smith, "Proceedings of the Aircraft Interior Materials/Fire Performance," April 4–5, 1989, Wichita State University, Kansas.
15. "Acousto-Ultrasonics, Theory and Application," edited by J. C. Duke, Jr. Plenum, N.Y. 1988.
16. N. E. Tornberg, SAMPE 33, 1738 (1988).
17. M. A. Hamstad, Exp. Mech. 26(1), 7 (1986).
18. J. Awerbach, M. R. Gorman and M. Madhudar, Mater. Eval. 43(6), 754 (1985).
19. Materials Engineering, pp. 38–41, May 1987.
20. S. A. St. Germain and M. P. Wai, SAMPE 34, 2141 (1989).
21. J. J. Singh, SAMPE 33, 407 and 1751 (1988).
22. "Size Exclusion Chromatography," Edited by B. J. Hunt and S. R. Holding, Routledge & Kegan Paul (New York) 1989.

5 Composite Interphase

5.1 Introduction

The high performance characteristics of advanced composite materials depend not only on the physical properties of fiber and matrix but also upon the interfacial region that exists between these dissimilar components. This interfacial region or interphase that is intermediate between the fiber phase and the matrix phase provides an important function in overall composite performance [1]. The interphase facilitates stress transfer from the moderately weak matrix resin to the high strength fiber and protects the fiber from environmental degradation. Stress can occur from differences in thermal expansion coefficients of fiber/matrix, cure shrinkage of thermoset matrices, or crystallization of thermoplastic matrices. Moreover the interphase can affect the matrix resin by preferential absorption of reactive monomer, oligomer or polymer or facilitate nucleation of crystalline high performance thermoplastic matrix.

Besides fiber/matrix interfacial region, knowledge of the chemical and physical mature of the interphase is necessary and requires an analysis of fiber surface. This is particularly important for carbon fibers since the carbon fiber surface possesses a number of functional groups (hydroxyl, carbonyl, carboxylic) and residual elements. Carbon fiber surface functionality has been detailed by XPS technique [2]. Carbon fiber analyses of surface defects (microporosity, cleavage cracks, crystallite boundaries, foreign inclusions or impurities and fracture-inducing flaws) and methods to remove or remedy these defects are equally important for optimum composite properties [3]. Skin-core effects of organic high performance fibers (aramid, PBO and PBT) cause weak locations and fiber defects and analoguous to "sheath-core" description of carbon fiber [3a].

Hence the fiber/matrix interphase, fiber surface effects and the interfacial bond (chemical and/or physical) are the major contributors to a formal understanding of the composite interphase.

5.2 Interphase Development

5.2.1 Chemical/Physical Bond

The development of a chemical or physical bond in the interfacial region can provide the necessary adhesion between fiber and matrix. A variety of modifiers can be used in improving the adhesion between these two dissimilar surfaces. These modifiers can be sizing agents, coupling agents, or surface modifiers.

A sizing agent is used to treat the fiber during manufacture and protect the fibers against abrasion during handling. Sizing agents are quite common for brittle fibers such as Glass and Carbon Fibers and consist of polyvinyl acetate, polyvinyl alcohol, starch and other natural and synthetic polymers.

Recently a high temperature glass fiber known as S-2 Glass HT fiber was introduced by Owens Corning for use with high temperature thermoplastic resins such as polyimides, PEEK, PEI, PPS, and liquid crystal polymers. The S-2 Glass HT fiber with a high temperature sizing agent survives processing temperatures up to 415 °C while the customary S-2 Glass was limited to 175 °C, representing the usual temperature range for epoxy resins. This latter temperature limitation is attributable to the glass size and not due to limitations in the glass. The primary function of the sizing agent is to prevent surface damage and breakage of the fibers during manufacture and handling. Protection is clearly important during weaving operations.

Coupling agents are known to promote fiber-matrix adhesion. They are bifunctional molecules with one functionality for reaction with the fiber surface and the second functionality capable of reacting with another material or the matrix. The coupling agent interacts chemically or physically with both the fiber surface and the polymer matrix. The most common type of coupling agents, especially for glass, is the silane coupling agent. Extensive work by Plueddemann has established silane coupling agent as the material of choice for many glass filled composites [4]. Various spectroscopic techniques have been used to characterize the nature of the bond formed between the coupling agent and fiber/matrix and include FTIR [5] Laser Raman Spectroscopy [6] and XPS [7], see Chap. 4 for Analysis/Testing.

Another class of materials are surface modifiers whose mode of adhesion is not completely understood. These include organotitanates, zirconates, aluminates and zircoaluminates [1]. These materials appear to aid in processing of composite components (plasticization, dispersing or lubricity) by an apparent lowering of surface energy of the substrate filler.

5.2.2 Surface Effects

5.2.2.1 Fiber

Studies directed to surface effects of high strength fibers (aramid, UHMWPE, S Glass and Carbon Fiber) are intensive. Some surface effects/treatments are proprietary and not disclosed by fiber manufacturers. The characterization of glass fiber surface has been instrumental in developing sizing agents and coupling agents that have optimized glass composite properties. Recently DiBenedetto [8] examined surface treatment of glass fibers with a variety of organosilane compounds and postulated that the interphase or an adsorbed layer is relatively short range or no more than 100–200 nm.

Surface modification of aramid by plasma treatment [9] improved bond strength of the composite. Takayanagi [10] modified aramid fiber surface by introducing functional groups via NaH/DMSO. The mechanical properties of the composites containing the functionalized fibers were greatly improved. Coating of aramid filaments with selected polyfunctional aziridines has a beneficial effect on ILSS of polyester matrix composites [11]. It is also reported that an epoxy or a polyvinyl alcohol surface size can be applied to aramid for improved adhesion [12].

A diversity of methods has been used to treat the Carbon Fiber surface. Methods include liquid and gaseous oxidizing agents, plasmas, chemical coupling agents, vapor deposition and electropolymerization. Surface oxidation resulting in acidic groups, carboxyl, lactone or hydroxyl groups are identified by XPS [13]. These special techniques are beyond the use of sizes which fiber manufacturers apply to carbon fiber. Common sizes include polyvinyl alcohol, epoxy, polyamide, and titanate coupling agents [14].

The relative inertness of UHMWPE makes it difficult to develop adhesion between the fiber and the matrix resin. Plasma treatment of UHMWPE fibers [15] causes some reduction in fiber molecular weight. However improved composite properties are obtained and compare favorably to untreated fiber composite properties.

5.2.2.2 Matrix Resin

The matrix resin is also susceptible to surface effects due to polymer morphology. The development of an interphase region can be facilitated by polymer matrix morphology particularly thermoplastic resins due to *1* polymer mobility imposed by fiber surface, or *2* polymer structure variations of bulk interior structure versus surface structure, *3* polymer morphology in the interphase, *4* the level of crystalline/amorphous nature of the thermoplastic resin. High performance semi-crystalline resins such as PEEK and PPS are examples of matrix resins with different nucleating effects in composites (Chapter 10, Thermoplastic Composites). With semi-crystalline polymers, nucleation densities and crystallization rates are important parameters that affect polymer morphology and influence interfacial adhesion and composite mechanical properties. Recently PPS composite data suggest [*16*] improved polymer toughness, morphological structure and interfacial adhesion contribute to improved laminate mechanical performance, better hot/wet behavior and substantial resistance to microcracking initiation.

Although most examples of resins influencing the interphase are known to be thermoplastic in structure, there are some reports of selected thermoset resins which also influence the interphase region. Some epoxy resins have an effect on the interphase due to changes in cure kinetics as a result of preferential absorption of curing agents such as polyamide curing agent by glass surface [*8*] or amine curing agent by carbon fiber surface [*17*].

5.3 Characteristics of the Interphase

Theoretical models [1, 18] of the interphase have been described in the literature. Maurer [19] considered the interphase as an "interlayer shell" that surrounds each particle in a discontinuous phase. By using ethylene vinyl acetate copolymer for the interphase and glass beads filled with styrene acrylonitrile resin, measured properties of individual phases allowed prediction of composite properties.

Physical properties of the interphase and their relationships to final composite properties have been proposed [1]. Depending on whether favorable composite strength or composite toughness is desired, a rigid interphase would provide the highest ultimate composite strength while a flexible interphase would promote composite toughness.

The concept of a "graded modulus" interphase has been examined by Kardos [20]. The concept consists of a material with an intermediate modulus between that of the fiber and matrix and reduces the stress at the interphase. Kardos [20] demonstrated that the preferential crystallization of polycarbonate near short graphite fibers resulted in higher modulus, higher strength composite. An alternative method of reducing stress concentrations at the interphase between brittle matrix resin and glass fibers is through the use of a ductile, energy absorbing material. This concept was originally suggested by Lavengood and Michno [21]. Kardos [22] extended the postulate by initially coating glass fiber with a flexible epoxy followed by filament winding with a brittle epoxy. Besides improved mechanical properties attributable to a local deformation mechanism that reduces interfacial stress concentrations, other variables beyond the benefit of interphase contributed to enhanced composite properties.

5.4 Interfacial Bond Strength

The determination of the interfacial bond strength is of considerable importance in assessing whether sufficient adhesion develops at the fiber/matrix interphase. An Interfacial Testing System based on an optical microscope has been reported [23]. The composite specimen is placed on the microscope with the fibers aligned perpendicular to the surface. A probe is pressed against the specimen until debonding is noticed. Fiber debonding results in a load drop which is sensed by a load cell attached to the specimen holder. Interfacial Shear Strengths (IFSS) are determined and allow comparisons among similar materials. The authors examined two types of carbon fibers with a cyanate resin matrix. The IFSS values allowed them to distinguish between which carbon fiber material could be improved in toughness with the same cyanate resin matrix. This optical microscope technique is similar to IFSS calculation [24] by embedding a single fiber within a matrix and stressing the matrix until the fiber breaks into several segments. The length at the fiber breaks fit a Weibull distribution from which the critical length and IFSS can be calculated.

The IFSS technique is applicable to a single fiber embedded in thermoset matrices but is quite difficult to apply to high performance thermoplastic matrices. A new technique [25] which has been used successfully in testing ceramic composites has been modified for use in determining the bond strength at the interface of PPS/Glass fibers. Values of 5.55–7.88 MPa were obtained by electrical resistance measurements. Fiber/matrix debonding, initiation of microcracks at the interface and crack growth are all monitored by the increase in resistance of a gold film that was sputtered onto the composite surface.

An interesting analytical method for determining the interface and/or interply strength for selective composites based on epoxy/carbon fiber or PPS/carbon fiber by using a thermoacoustic technique has been reported by Wu [25a]. Preliminary data suggest that a correlation exists between acoustic output and the interfacial strength; samples with weaker interfaces yield more acoustic output than those with a stronger interface when subjected to local heating. Further work is in progress to utilize advanced analysis schemes to determine the source of acoustic energy by considering materials and stress.

5.5 Future Efforts

With the advent of new instrumentation such as Scanning Tunneling Microscope (STM) [26] and Magnetic Resonance Imaging (MRI) [27], studies directed to the fiber/matrix interphase will provide a better understanding of the dynamics of the interphase and possible correlation of interfacial strength/properties with interfacial adhesion.

Surfaces can be imaged with the STM at atomic resolution as little as 1 Å with the creation of three dimensional images that illustrate actual atomic position and defects. The STM is useful not only for analytical purposes but also for surface modification. The STM can modify surfaces on an atomic scale and image the effect immediately.

With MRI, new information can be generated as it relates to the interphase and interphase properties. MRI has been effective in identifying porosity in fibers, ceramics and the distribution of water in composites. This latter capability will lead to a better understanding of the role of the interphase in a humid environment.

Continued efforts in improving the determination of interfacial bond strengths as well as Interfacial Shear Strengths will be beneficial in the design of an optimum interphase for ultimate composite performance.

5.6 References

1. J. D. Miller and H. Ishida, "Adhesive-Adherend Interface and Interphase", in "Fundamentals of Adhesion", L. H. Lee, editor, Plenum, N.Y. (in press).
2. T. Takahagi and A. Ishitani, in "Molecular Characterization of Composite Interfaces," H. Ishida and G. Kumar, editors, Plenum Press, N.Y. (1985).
3. I. L. Kalnin and H. Jager, in "Carbon Fibers and Their Composites," E. Fitzer, editor, Springer, Berlin Heideberg New York (1985).

3a. Ibid. p. 64.

4. E. P. Plueddemann, "Silane Coupling Agents," Plenum, N.Y. 1982.
5. H. Ishida and J. L. Koenig, J. Col. Intf. Sci 64(3), 555 (1978).
6. J. L. Koenig, "Chemically Modified Surfaces," Volume 1, D. E. Leyden, editor, Gordon and Breach, N.Y. (1986).
7. L. V. Phillips and A. M. Hercules, "Chemically Modified Surfaces," D. E. Leyden, editor, Gordon and Breach, N.Y. (1986).
8. A. T. DeBenedetto and P. J. Lex, Poly. Eng and Sci. 29(8), 543 (1989).
9. M. R. Wertheimer and H. P. Schreiber, J. Appl. Poly. Sci., 26, 2087 (1981).
10. M. Takayanagi, T. Kajiyama and T. Katayose, ibid., 27, 3903 (1982).
11. F. M. Logulla, Sr. and Y-T. Wu, U.S. 4,418,164 (11/29/83) Dupont.
12. K. K. Chawla, "Composite Materials," p. 97, Springer, Berlin Heidelberg New York (1987).
13. I. L. Kalnin and H. Jager, Chapter 1 in "Carbon Fibers and Their Composites," E. Fitzer, editor, Springer, Berlin Heidelberg New York (1985) p. 66.
14. Reference 12, p. 157.
15. H. X. Nguyen, G. C. Weedon and H. W. Chang, SAMPE 34, 1603 (1989).
16. R. L. Hagenson, D. F. Register and D. A. Soules, ibid., 34, 2255 (1989).
17. L. T. Drzal, M. J. Rich, M. F. Koenig, P. F. Lloyd, J. Adhesion 16, 133 (1983).
18. P. S. Theocaris, G. Spathis and B. Kefales, Colloid & Polym. Sci., 260, 837 (1982).
19. Reference 1, p. 27.
20. F. S. Cheng, J. L. Kardos and T. L. Tolbert, SPE Journal 26, 62 (1970); J. L. Kardos, J. Adhesion, 4(1), (1972).
21. R. E. Lavengood and M. J. Michno, Jr. Proc. Div. Techn. Conf. Eng. Prop. Structure Div., SPE, 127 (1975).
22. J. L. Kardos, Chem Tech, July, p. 430 (1984).
23. D. L. Caldwell and D. A. Jarvie, SAMPE 33, 1268 (1988).
24. W. A. Fraser, F. H. Ancker, A. T. DeBenedetto and B. Elbirli, Poly. Comp., 4, 238 (1983).
25. C. T. Necker, V. Krishnamurthy, M. W. Barsoum and I. L. Kamel, SAMPE 34, 2161 (1989).

25a. W-L. Wu, SAMPE J., 26(2), 11 (1990).

26. J. P. Rabe, Angew. Chem. Int. Ed. Engl., 28(1), 117 (1989).
27. J. L. Ackerman, Poly Prep., 29(1), 88 (1988).

6 Composite Mechanical Properties

6.1 Introduction

Advanced composite materials, e.g. carbon fiber reinforced polymeric matrices, have found wide-scale application in the sporting goods and aerospace markets. Application of composites to sporting goods is based on the material's lightness, strength, weather resistance, and dimensional stability [1]. The prime driver for aerospace composites applications is the ability to reduce mass, hence, save weight, by tailoring designs and materials to achieve structure which is resistant to failure under in-service loads and environments [2].

The advanced materials/composites industry (both materials suppliers and end-users) has grown to understand key composite mechanical properties and why they are important. Composites afford the opportunity to tailor both mechanical and physical properties to achieve specific end-use requirements. To effectively tailor properties, the structure-property relationships between the matrix resin, reinforcement fiber, and fiber/matrix interface must be understood. To effectively achieve optimal weight structure, designers must understand how much of a potential strength or stiffness improvement can be exploited when realistic design criteria are imposed. Real life design criteria often involve stability under compressive loading, the effect of stress (or strain) concentrations due to cut-outs and notches, and the effect of impact induced damage [2, 3].

Fundamental lamina and laminate mechanical properties and test methods will be discussed. The compilation of these properties results in a data base to allow materials selection and hardware design to meet end-use requirements. Advanced composites test methods generally follow the recommendations of the American Society for Testing and Materials (ASTM) in the USA and the Composite Research Advisory Group (CRAG) in the UK. Composite property data bases, which employ the above test methods, are being generated under a number of cooperative efforts including MIL-HDBK-17 and the test methods task force of SACMA, Suppliers of Advanced Composite Materials Association in the USA.

6.2 Lamina Properties

The basic building block for all continuous fiber reinforced composites is the specially orthotropic lamina as shown in Fig. 1. The terminology longitudinal

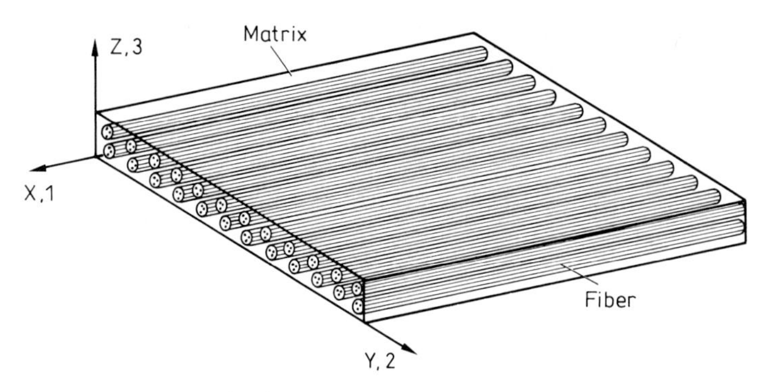

Fig. 1. Composite lamina

Table 1. Lamina mechanical and physical properties

Property	Symbols	Units	ASTM Test Method
Moduli		GPa	
Axial or longitudinal	E_a, E_{11}		ASTM D3039
Transverse	E_t, E_{22}		ASTM D3039
Shear	G_a, G_{12}		ASTM D3039
Axial Poissons' ratio	ν_a, ν_{12}	(Dimensionless)	ASTM D3039
Transverse Poisson's ratio	ν_t, ν_{21}	(Dimensionless)	ASTM D3039
Strength[1]		MPa	
Axial tensile	σ_a^{tu}		ASTM D3039
Axial compressive	σ_a^{cu}		ASTM D695
Shear	τ_a^u		ASTM D2344
Transverse tensile	σ_t^{tu}		ASTM D3039
Transverse compressive	σ_t^{cu}		ASTM D695
Thermal expansion		10^{-6}/°C	
Axial	α_a, α_{11}		ASTM E831
Transverse	α_t, α_{22}		ASTM E831
Thermal Conductivity		W/mK	
Axial	K_a, K_{11}		ASTM E1225
Transverse	K_t, K_{22}		ASTM E1225
Specific heat	C	kJ/kg K	ASTM E968
Density	ρ	kg/m^3	ASTM D792

[1] Ultimate strain in the axial and transverse direction, E_a and E_t expression in % are often more important than ultimate strength.

or axial direction denotes the fiber axis while the term transverse direction references the in-plane and out-of-plane axes at right angles to the fiber axis.

The fiber axis is often designated as the x or 1 axis in a right handed coordinate system while the in-plane transverse direction is designated as the y or 2 axis and the out-of-plane transverse direction is labeled the z or 3 axis. A basic lamina is a single ply of unidirectional prepreg tape.

The mechanical and physical properties most often measured to characterize lamina performance are listed in Table 1. Also listed in Table I are frequently used property nomenclature, units, and ASTM test methods. In addition to the ultimate composite strengths in the axial and transverse directions the ultimate strains at failure (expressed in %) are also very important in the design of structures. Strain allowables as opposed to strength allowables may be the limiting design criteria.

A discipline within continuum or solid mechanics, known as "micromechanics" has generated predictive equations that are able to relate the fundamental properties of the constituent fibers and matrices to the lamina properties. These most important "structures-properties" relationships are described by Ashton [4] and McCullough [5]. The properties of laminates (macromechanics) are predicted using laminated plate theory [6] and the basic lamina

Table 2. Typical lamina properties PAN and pitch composites (Standard epoxy $V_f = 0.60$)

Fiber	T300	T650-42	P-75	P-120
Moduli				
E_a (GPa)	140	170	310	510
E_t (GPa)	13	13	6.6	5.9
G_a (GPa)	5.5	5.5	4.1	4.8
ν_a	0.3	0.3	0.29	0.29
ν_t	0.02	0.02	0.006	0.003
Strength				
σ_a^{tu}, (MPa)	1800	2400	900	1100
σ_a^{cu}	1730	1730	410	338
τ_a^u	104	104	220	110
σ_t^{tu}	55	55	28	21
σ_t^{cu}	240	240	69	48
Thermal expansion				
α_a, (10^{-6}/°C)	−.18	−.36	−1.3	−1.4
α_t, (10^{-6}/°C)	23.4	27.0	36.2	37.8
Thermal conductivity				
K_a, (W/mK)	5.1	9	110	350
K_t, (W/mK)	—	—	1.1	1.1
Specific heat				
C, (kJ/kgK)	1340	1340	1340	1340
Density				
ρ, (kg/m³)	1570	1600	1700	1900

properties. Numerous books have been published on the subjects of both micro- and macro-mechanics.

Details of micro- and macro-mechanics are beyond the scope of this publication. However, the basic influence of fiber, matrix resin, and interfacial properties on specific lamina and laminate properties will be discussed in Chapter 7.

Table 2 summarizes baseline lamina properties for a standard epoxy matrix resin reinforced at 60% fiber volume for standard modulus carbon fiber (Thornel T300), intermediate modulus carbon fiber (Thornel T650-42), high modulus pitch fiber Thornel (P-75) and ultra high modulus pitch fiber (Thornel P-120). Examination of these composite lamina properties discloses that transverse properties are much lower than axial properties, that lamina tensile strengths exceed compressive strengths, and that axial moduli for high and ultra high modulus pitch-fiber based composites far exceed those achievable with PAN-based fibers. Shear strengths must also be critically examined when considering end-use requirements.

Table 3. Critical laminate properties

Test	Ply Orientation	Specimen Configuration, mm	Required Data	Units of Measure
Tension	$(0)_8$	12.7 by 228.6 tabbed	Strength	MPa
Tension	$[\pm 45]_{3s}$	25.4 by 228.6 tabbed	Modulus	GPa
			Strength	MPa
			Shear Modulus	MPa
Compression	$(0)_8$	12.7 by 80 tabbed	Strength	MPa
			Modulus	GPa
Open-hole tension	$[45/0/-45/90]_{2s}$	38.1 by 304.8 6.35 diameter hole	Strength	MPa
Open-hole compression	$[45/0/-45/90]_{2s}$	38.1 by 254 6.35 diameter hole	Strength	MPa
Compression after impact	$[45/0/-45/90]_{6s}$	12.7 by 254	Strength	MPa
Compression interlaminar shear	$[45/0/-45/90]_{6s}$	12.7 by 80	Strength	MPa
Edge delamination tension test	$[(\pm 30)_2/90/90]_s$	38.1 by 254	Interlaminar fracture toughness	MPa

6.3 Laminate Properties

As was mentioned earlier, the lamina properties may not be the determining factor in composite structural performance. Realistic design criteria must consider laminate properties in tension and compression in the presence of stress risers e.g. a hole, and residual properties after impact damage and at elevated temperature in the presence of hostile environments. Laminate strength in the presence of both loaded and unloaded holes and notches is particularly important for thin composites structure e.g. helicopter fuselage skins. Residual compressive strength in the presence of impact induced damage is important for thick composite structure e.g. airplane wing skins.

The need for reliable design data for laminates with holes and impact damage has resulted in the evolution of a series of critical laminate tests as documented in Chapters 7 and 9. Specific emphasis has been placed on these laminate tests with the evolution of more damage tolerant thermosetting matrices and thermoplastic matrices. Key tests include open-hole tension and compression (OHT and OHC) and measures of compression-strength-after-impact (CAI) or post-impact-compression (PIC).

Critical tests and laminate configurations are tabulated in Table 3. Typical values for standard epoxies, damage tolerant epoxies, and amorphous thermoplastics are listed in Table 4.

Table 4. Typical laminate properties (intermediate modulus carbon fiber)

Test	Std. Epoxy	Damage Tolerant Epoxy	Amorphous Thermoplastic
0° Tension Strength (MPa)	2400	2400	2400
± 45° Tension			
Modulus, (GPa)	21	18	19
Strength, (MPa)	186	255	248
Shear modulus, (GPa)	5.6	6.3	4.8
0° Compression			
Strength, (MPa)	1700	1700	1300
Modulus, (GPa)	152	152	152
OHT Strength (MPa)	480	449	483
OHC Strength, (MPa)	310	304	269
CAI, (MPa) @ 6700 J/m	160	311	304
Compression interlaminar, (MPa)	90	76	69
EDS, (MPa)	214	290	311

Approximate fiber volume = 60%.

6.4 Summary

A well established data base on composite lamina is the building block for structural applications of composites. The materials suppliers can use micromechanical analysis in conjunction with constituent properties to predict improvements in lamina performance. End-users of composites can use macromechanical analysis to predict properties of structural laminates based on lamina properties. However, usable composite structural properties must account for the presence of cutouts, holes, and impact induced damage. Chapter 7 provides composite structure property relationships. Chapter 9 examines critical laminate properties needed to achieve balanced composite performance at elevated temperatures in the presence of both hostile environments and either impact damage or machined holes needed to assemble subcomponents into finished structure.

6.5 References

1. R. Pinzelli and D. Vanthier, Kunststoffe 77, (1987) 9, 876.
2. R. F. Simenz, Phil. Trans. R. Soc. Land. A322, 15, 1987.
3. C. Rodgers, W. Chan and J. Martin, Proc. 43rd Ann. Forum American Helicopter Soc., 597, 1987.
4. J. E. Ashton, J. C. Halpin and P. H. Petit, Primer on Composite Materials: Analysis, Technomic, 1969.
5. R. L. McCullough, Concepts of FiberResin Composites, Marcel Dekker, Inc., 1971.
6. J. E. Ashton and J. M. Whitney, Theory of Laminated Plates, Technomic, 1970.

7 Composite Structure – Property Guidelines

7.1 Introduction

The objective of this chapter is to document the effect of constituent properties on composite properties. Detailed micro-mechanics formulae will not be discussed. However, the functional relationships will be analyzed to provide guidelines concerning which constituent property should be modified to effect a change in a particular composite property. The constituent properties of interest for fibers appear in Table 1, Chap. 3. Matrix properties of interest appear in Tables 4 and 9, Chap. 2. Interfacial effects discussed in Chapter 5 will be amplified as they relate to composite properties.

Discussions which focus on the selection of matrix resins to improve composite properties have been presented by Serafini [1] and McCullough [2]. Trends in composite performance in the presence of aqueous or solvent environments at room or elevated temperature are discussed by Chamis [3, 4] and Collings [5]. Additional composite structure property functional relationships can be found in [6] and [7]. Analyses of composite moduli, composite strength and the influence of variables that relate to these characteristics such as matrix resin, fiber/fiber volume, interface/interphase, and method of testing are described. Tests such as edge delamination strength and $\pm 45^\circ$ tension provide a measure of balance of composite system performance, i.e., toughness and in-service environmental performance.

7.2 Composite Moduli

Micro-mechanical relationships [6] to predict composite moduli are the most highly developed and most reliable in the generation of data. Axial composite moduli depend directly (in a linear relationship) on the fiber modulus and the volume of fiber in the composite. Transverse composite modulus is inversely dependent on the fiber modulus and directly dependent on the matrix modulus. Composite shear modulus and Poisson's ratio depend directly on the matrix modulus. All the above properties are dependent on the fiber volume fraction and the geometry of reinforcement. A complete discussion of micro-mechanical equations and structure property relationships is presented in [7].

In general, the highest composite axial modulus will result when the highest modulus fiber is employed at the highest fiber volume fraction. The highest transverse composite modulus and composite shear modulus will result from the highest modulus matrix resin. Implicit in these discussions on composite structure property guidelines is the assumption that the laminates being characterized are of high quality. Specifically, the fiber and matrix resin are uniformly distributed, void level is low (generally less than 1 volume %), and that residual thermal strains and microcracking induced by the cure/consolidation cycle are minimal. Nondestructive evaluation of cured laminates is essential to ensure quality (Chapter 4). Furthermore, test coupons must be machined to minimize edge damage and ensure alignment of the fibers in the desired direction. Test technique is also very important and requires accurately calibrated testing machines, correctly installed extensometers and/or strain gauges, and proper alignment of the test coupons in the testing machine grips. Detailed discussion of proper test techniques and the effect of improper testing in measured composite properties are described in [8–15].

7.3 Composite Strength

Table 1 lists basic composite strengths, the constituent property which strongly effects each strength, and the ASTM test techniques most often used to characterize strength.

Axial tensile strength is directly dependent on the ultimate strength of the fiber, the fiber volume fraction, the matrix modulus, and ultimate strain or matrix ductility. For a given matrix ultimate strain, the composite tensile strength will increase linearly with the matrix modulus. However, the interfacial bond between the fiber and the matrix must be optimal. Too low an interfacial shear strength leads to premature debonding while too high an interfacial strength can result in a brittle longitudinal or axial splitting failure mode. Axial tensile strength is also extremely sensitive to fiber alignment. Fiber misalignment as little as 1/4 to 1/2 a degree off-axis can reduce apparent composite tensile strength by 25 to 50%. In general, to achieve high composite axial tensile strength with a fixed interfacial shear strength, it is desirable to have the highest strength fiber and a ductile, high ultimate strain matrix resin.

Axial compressive strength[1] is highly dependent on the failure mode, which is correspondingly dependent on the constituent properties. The highest strengths will result from composites based on fibers with high inherent compressive strength, with excellent alignment in a matrix resin with a high tensile (and hence shear) modulus. This combination forces a compressive failure of the fibers. Other dominate failure modes which result in less than optimal composite compressive strength are symmetric or nonsymmetric buckling predominately caused by a low matrix modulus and debonding or interfacial shear failure caused by a weak interfacial bond.

[1] Axial compressive strength is discussed more thoroughly in Chap. 8.

Table 1. Composite strength: Influence of constituents

Composite strength	Dominant constituent property			Comments	ASTM test method
	Fiber	Matrix	Interface		
Axial tensile	Ultimate strength & Strain	Modulus & Ductility	Value vs matrix shear strength	Most influenced by fiber very sensitive to ductile low modulus matrix and very high interfacial bond.	D3039
Axial compressive	Strength	Modulus	Value vs matrix shear strength	Failure mode key i.e fiber failure either compression or local buckling. High matrix modulus desired	D695
Transverse tensile	—	Ductility Modulus Toughness	Value vs matrix shear strength	Very sensitive to flaws, stress risers. Ductile matrix desired. High G_{1c}.	D3039
Transverse compressive	—	Ductility Modulus Toughness	Value vs matrix shear strength	Ductile matrix with high G_{1c} desired.	D695
Intralaminar shear	—	Modulus Ductility Toughness	High value desired	Ductile matrix with high interfacial bond desired. Flaw sensitive.	D3518
Interlaminar shear	—	Modulus Ductility Toughness	High value desired	Flaw (void) sensitive. Ductile matrix with high interface bond desired.	D2344

Detailed observation of composite axial compression failure mode is particularly important for different types of fiber reinforcement. PAN based fiber composites often fail by a matrix, interfacial, or instability dominated mode indicating that optimal compressive strength, i.e. based on fiber compressive strength, has not been achieved. Additional composite compressive strength may result from increases in e.g. the matrix modulus. On the other hand, the inherent micro structure of pitch carbon fibers leads to composite compressive failure mode triggered by fiber failure via slip on critical crystallographic planes. This failure mode would indicate that limiting composite compressive strength has been achieved. A detailed discussion of the effect of fiber type on composite compressive strength appears in [16]. A high value of axial composite compressive strength is of particular importance in aircraft design. Due to the alternate tensile and compressive in service loading, nearly 80% of aircraft structure is critically loaded in compression. Retention of high axial compressive strength, particularly in an elevated temperature/wet state is critical to structural performance. Most matrix resins, damage tolerant matrices in particular, experience a reduction in modulus under elevated temperature, wet conditions and hence, composite compressive strength under in-service conditions must be well characterized.

Composite transverse tensile and compressive strength as well as inter and intra laminar shear strengths are directly dependent on the matrix modulus and the ultimate matrix strain or ductility. In general, the transverse and shear strengths will increase nearly linearly with increasing matrix modulus. However, a high matrix ultimate strain is desired to achieve a composite highly resistant to the inherent strain concentrations and thermal strains induced by reinforcement and cure/consolidation. Transverse and shear properties will be particularly sensitive to the presence of flaws e.g. voids. Hence, void volume fraction should be minimized to maximize transverse and shear strengths.

The more desirable tests for intralaminar shear strengths (rod torsion and off axis tension) and for interlaminar shear (short beam shear) are highly dependent on specimen geometry, e.g. rod diameter, span-to-depth ratio etc. as well as void content. A better measure of relative interfacial shear strength is edge delamination strength which will be discussed in some detail.

An excellent assessment, based on composite data, for the effect of constituent changes on composite properties can be achieved by reviewing Ref. 16. It presents neat resin properties for ten matrices, fiber properties on four different carbon fibers, and a very complete composite property data base.

7.4 Edge Delamination Strength

The edge delamination strength is derived from the on-set of matrix shear failure, interfacial shear failure, or interply debonding in a laminate tensile test. The test coupon and pertinent ply lay-up and dimensions are shown in Fig. 1, and discussed in detail in Refs. 17, 18.

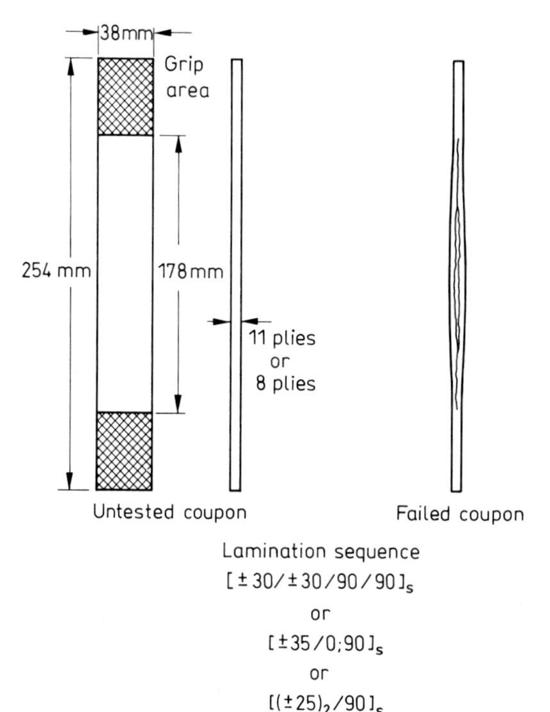

Fig. 1. Edge delamination test laminate

This test coupon is easy to fabricate and requires no special end tabbing. Hence, it is a convenient method for deriving relative composite performance for various combinations of fiber, matrix resin, and fiber interfacial or shear treatments. Testing is conducted in tension using either a screw-driven or servo-hydraulic test machine in the stroke or strain control mode. Tensile loading is continued until visible edge delamination is detected. Painting or chalking the laminate edge with white powder assists in accurate detection of delamination. The load (or stress) and strain at the on-set of delamination are recorded, as shown schematically in Fig. 2.

The prime factor for delamination is the poorly matched Poisson's ratio between the transverse inner plies and the external angle plies. The transverse plies deform (strain) very little relative to the outer angle plies thus driving a delamination. The delamination may occur at the inter ply region, at the fiber resin interface, or in the matrix resin of the inner plies. The lowest value of any of the strengths such as interlaminar, matrix shear, or interfacial shear strength, will govern the failure mode, which is recorded for analysis of structure-property relationships and comparisons.

Knowledge of the specimen geometry, loads, moduli and strains allows one to calculate the edge delamination strength or interlaminar fracture toughness. Table 2 contrasts edge delamination strength results for a variety of fiber/matrix resin combinations.

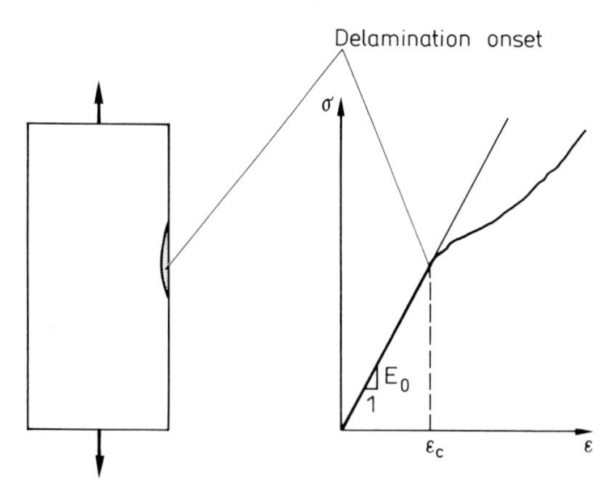

Fig. 2. Schematic onset of delamination

Table 2. Edge delamination strength (EDS)

Standard modulus carbon fiber	EDS, (MPa)
Brittle epoxy	138–207
Ductile epoxy	345–448
Brittle BMI	69–103
Ductile BMI	207–276
Intermediate modulus carbon fiber	
Brittle epoxy	103–172
Ductile epoxy	345–414
Brittle BMI	0–83
Ductile BMI	138–241
BMI (fixed ductile system)	
Carbon fiber no shear treatment	34
Carbon fiber standard shear treatment	138–172
Carbon fiber high shear treatment	138–207

The contrast and sensitivity of edge delamination strength is strongly related as constituent properties are varied. This efficient test method provides immediate feedback to guide development of optimized composite systems. Depending on the failure mode, it is readily apparent where modification is needed e.g. more fiber/matrix interfacial strength or more toughness in the matrix resin. Test results also give experienced designers an understanding of the most probable behavior of composite structure with cut-outs or holes. EDS test results also correlate well with compression strength-after-impact (CAI) or post impact compression (PIC). Specifically, cases are not known where high EDS values resulted in low CAI or PIC values, provided that laminate quality is comparable. Another laminate test which provides useful structure-property guidance is the $\pm 45°$ tensile test in which the shear modulus retention under hot/wet conditions is monitored.

The combination of EDS testing and $\pm 45°$ tensile testing hot/wet, provides a measure of the balance of composite system performance i.e. toughness and in-service environmental performance. In general structure property trends as measured by either EDS or $\pm 45°$ tension correlate very well.

7.5 $\pm 45°$ Tension

Like the EDS test coupon, the $\pm 45°$ tensile test coupon is a simple laminate which is easily fabricated and tested. The laminate is 8 plies thick, $[(\pm 45)_2]_s$, 25 mm wide, overall length is 300 mm, with a 125 mm gauge length. The long specimen length affords a large grip area, end tabs are not needed nor desirable since the high elongations of this specimen would make retention of bonded end tabs very difficult.

The critical tests to be conducted are room temperature dry tests and elevated temperature wet tests. Generally, modulus, ultimate strength, ultimate strain, and modulus retention are the properties, measured and/or calculated. Conditioning the sample in water is conducted by immersion for a fixed period of time in a water bath at a controlled temperature or by exposure to a controlled temperature and relative humidity until a particular weight gain or saturation is achieved.

Table 3 lists test results for composites based on: standard epoxy, damage tolerant epoxy, BMI (bismaleimides), and thermoplastic matrices with an intermediate modulus carbon fiber as the reinforcement.

Sensitivity is high and absolute values easily discriminate performance among matrix resins for a fixed fiber. Conversely the matrix resin could be fixed

Table 3. $\pm 45°$ Tensile test data[a]

	Strength (MPa)	Modulus (GPa)
Standard epoxy @ RTD[b]	186	21
DT epoxy @ RTD	255	18
BMI @ RTD	193	19
TP @ RTD	248	19
Standard epoxy @ 135 °C/Wet	90	12
DT epoxy @ 104 °C/Wet	124	12
BMI @ 190 °C/Wet	97	7
TP @ 149 °C/Wet	124	15

[a] Nominally 60% fiber volume.
[b] RTD = Room Temperature Dry.

and the effect of fiber variations and/or fiber shear treatment variants could be characterized; $\pm 45^\circ$ tensile test results provide a good measure of elevated temperature wet performance. The effects of other solvents or fluids besides water could also be effectively measured via the $\pm 45^\circ$ tensile test.

7.6 Summary

The tailoring of composite properties through variations in fiber type, matrix resin type, and interfacial shear strength affords the opportunity to optimize end-use structural performance. Axial, transverse, and shear properties depend on the constituent properties. Detailed micro-mechanical predictions are possible through empirical and theoretical relationships. Guidelines have been presented for key matrix properties to effect changes in axial and transverse tensile or compressive strength. Additionally, the use of the edge delamination tensile tests to elucidate interfacial shear performance has been described. The merits of $\pm 45^\circ$ tensile testing to gauge composite system performance at elevated temperature/wet conditions were also discussed.

7.7 References

1. C. C. Chamis, M. P. Hanson and T. T. Serafini, Modern Plastics, May 1973, p. 90.
2. R. L. McCullough, Concepts of Fiber Resin Composites, Marcel Dekker, New York, 1971.
3. C. C. Chamis and J. H. Sinclair, SAMPE Quarterly, Oct., 1982, p. 30.
4. C. C. Chamis, SAMPE Quarterly, April, 1984, p. 14.
5. T. A. Collings and D. E. W. Stone, Composites, Vol. 16, No. 4, Oct., 1985, p. 307.
6. J. E. Ashton, J. C. Halprin and P. H. Petit, Primer on Composite Materials Analysis, Technomic, 1969.
7. L. E. Neilsen, Mechanical Properties of Polymers and Composites, Vol. 1 & 2, Marcel Dekker, New York, 1974.
8. P. W. Manders, Sci. of Adv. Matl. & Proc. Engr. Series, Vol. 32, 1987.
9. S. B. Batdorf, J. Reinforced Plastics and Composites, Vol. 1 (April 1982), p. 153.
10. S. B. Batdorf, J. Reinforced Plastics and Composites, Vol. 1 (April 1982), p. 165.
11. M. P. Nemeth, C. T. Herakovich and D. Post, Composites Technology Review, Vol. 5, No. 2, 1983, p. 61.
12. G. M. Newaz, SAMPE Quarterly, Jan. 1984, p. 20.
13. G. M. Newaz, Matls. Engr., May 1984, p. 21.
14. J. C. Weidner, Proc. European SAMPE, 1983.
15. I. M. Kowalski, 31st SAMPE, 1986.
16. R. S. Zimmerman and D. F. Adams, NASA Contractor Report 181631, Grant NAG1277, Dec. 1988.
17. NASA Reference Publication 1092, 1983.
18. NASA Reference Publication 1142, 1985.

8 Composite Compressive Strength

8.1 Introduction

The importance of composite compressive strength is fortified by the wealth of published literature pertaining to this subject. A literature search disclosed more than 350 references. Approximately twentyfour references were identified as important developments and span the timeframe since 1981. Older references, although important, are not cited since the significance of the conclusions and validity of approach of the more recent works is given priority.

A review of the literature indicates that a concise, quantitatively accurate, micro-mechanics model of composite compressive strength does not yet exist. However, the various analytical models proposed and supporting experimental data help define the controlling damage mechanisms and directions to pursue in order to increase compressive strength.

The salient features of the important literature contributions will be discussed along with a review of various compression test methods commonly used. Composites performance as it relates to open-hole-compression testing (OHC) will be assessed. Compressive strength materials specification/design criteria which are being used to screen candidate materials form the basis for data bases to be reviewed.

8.2 Compression Strength/Structure Property Status

The basis for this summary is found in Refs. 1 to 24. It is generally accepted that fiber-reinforced composites can fail in compression (axial) with various failure modes depending on the constituent materials properties and also upon the presence and extent of defects. Governing failure modes in carbon fiber/epoxy (and other polymeric matrices) include: kinking (or shear crippling), pure compression failure of the fibers, or longitudinal splitting. It is understood that buckling or kink band formation is more evident with low modulus fibers (e.g., Aramid, Spectra) or softer polymers, whereas failure of composites based on stiffer fibers may be due to the actual compressive failure of the fiber itself. The term "softer" polymers refers to low modulus [less than 1720 MPa] high elongation or ductile polymers. Stiff fibers include PAN or pitch based fibers with an axial modulus in excess of 350 GPa.

While no quantitative model exists to determine which mode of failure will occur, it has been observed that most PAN based fibers fail at much higher failure strains with microbuckling or kinking failure modes, while higher modulus pitch fibers (e.g. Thornel P-75), have low failure strains with longitudinal splitting, shearing or pure compressive failure of the fibers themselves [24a, b]. Several microbuckling models have been proposed, and generally relate composite compressive strength to the shear modulus or yield strength of the matrix resin. The basis for these theories is the concept that the loss of lateral support from the matrix precipitates failure. Most of the models predict higher compressive strength than is actually observed. Much of the recent literature focuses on various methods to bring predicted and measured strengths more closely in agreement.

Factors taken into account in the newer models include fiber/matrix disbonding, fiber waviness, 3-dimensional buckling shapes, voids and inclusions. In most cases, when these factors are introduced, the problem is too complex to be modelled directly and empirical evaluation must be used. One can deduce from existing models that if compressive failure is initiated by micro-buckling or kink-band formation, increased compressive strength could occur via the following: the matrix shear modulus and yield strength are increased; the interfacial (fiber to matrix) shear bond strength is increased; fiber waviness and void or defect content are decreased.

If the governing failure mode is the actual compressive strength of the fiber, then obviously, increasing composite compressive strength requires increasing the intrinsic compressive strength of the fiber itself. Methods of achieving increased fiber compressive strength remain highly proprietary to fiber manufacturers.

8.3 Compression Testing Methods

The measured compressive strength of uni-directional composites is very sensitive to test technique and sample preparation. Compression testing methodology is reviewed and a historical perspective on several techniques is provided in [25–28]. The motivation for all test methods is to provide a configuration which ensures a composite compressive failure by eliminating the possibility of a gross buckling failure and/or crushing of the specimen ends. Hence, emphasis in compression testing must be on sample preparation (flat and parallel test coupon ends) and alignment fixtures to ensure the desired axial loading and constraint against premature specimen buckling.

The compression testing techniques and associated fixtures which have received most attention include: University of Dayton Research Institute (UDRI) sample; the Texaco test; Royal Aircraft Establishment (RAE) test; the Celanese test, the Illinois Institute of Technology Research Institute (IITR) test and fixture; the Northrop test (ASTM D-695-77) and tabbed Northrop test; and the sandwich beam test.

The latter two test techniques are the most widely used. The Northrop or ASTM D-695 test fixture and coupon is shown in Fig. 1 in an untabbed

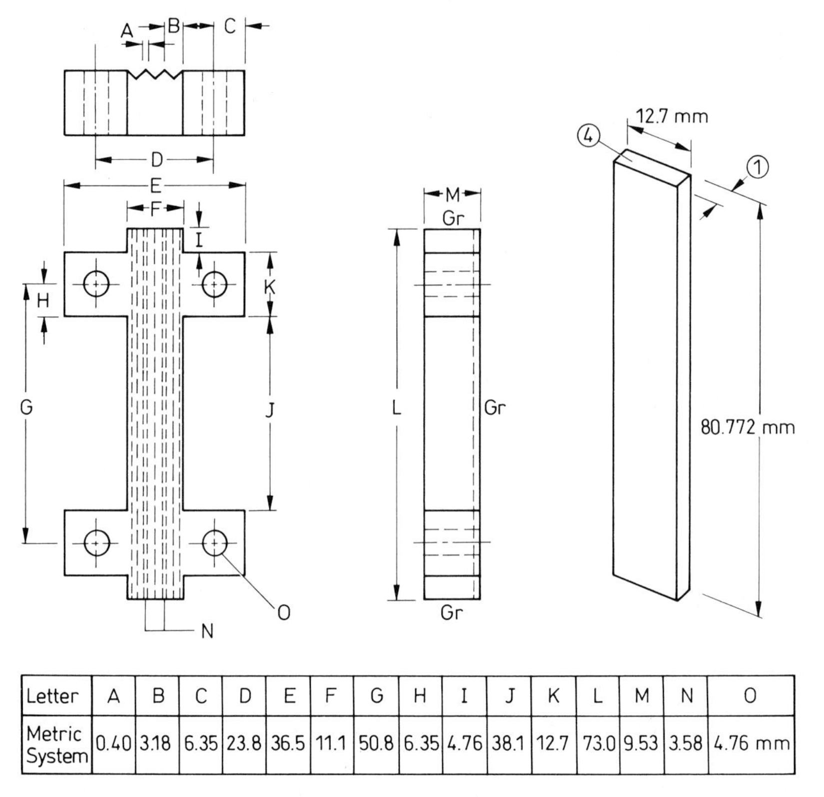

Letter	A	B	C	D	E	F	G	H	I	J	K	L	M	N	O
Metric System	0.40	3.18	6.35	23.8	36.5	11.1	50.8	6.35	4.76	38.1	12.7	73.0	9.53	3.58	4.76 mm

Fig. 1. Northrop compression test: *1* Thickness from > 1 mm. *4* Grind end flat and parallel

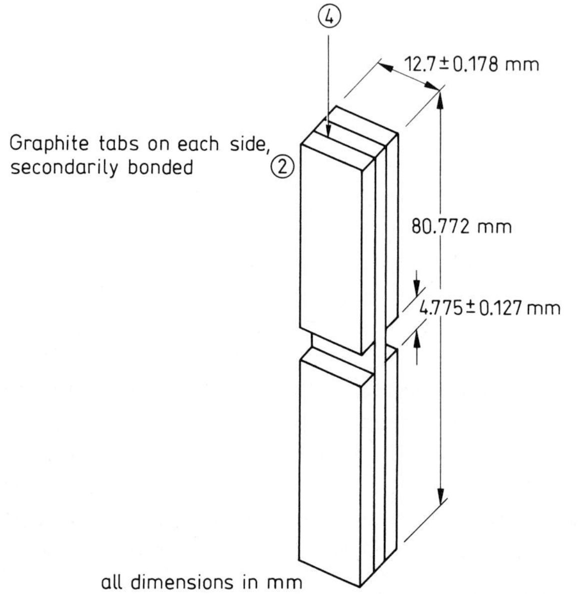

Fig. 2. Tabbed northrop compression specimen: *2* graphite tabs from same composite as the specimen. *4* grind ends flat and parallel

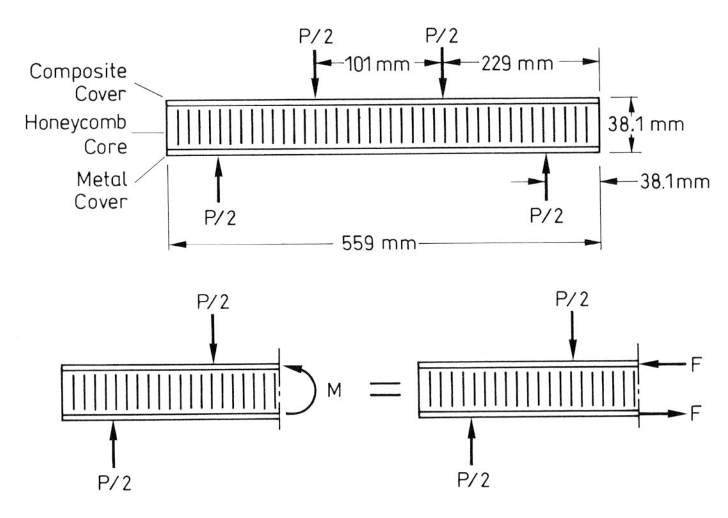

Fig. 3. Sandwich beam compression test

configuration and in Fig. 2 for the tabbed configuration. The tabs assist in preventing premature failure via end-crushing. A typical sandwich beam sample and 4-point loading in flexure is shown in Fig. 3. In this sample, the metal skin on the tensile side of the coupon is an effective means of forcing a compressive failure on the upper composite skin which is loaded in compression. A great deal of technique is required in preparing the sandwich beam sample, particularly in ensuring effective bonding between the honeycomb core and the face sheets or skins. Sound test coupon preparation and axial loading of properly fixtured samples ensures reliable compressive failure of the composite.

8.4 Axial Compressive Strength Requirements and Properties

Compressive strength is extremely important particularly in aircraft applications where at any given moment up to 80% of the critical structure can sustain compressive loading. Requirements for composite compressive strength (and other properties) are heavily dependent on the application e.g. in-service environments and stress levels, either or both being quite different for a civil aircraft and a supersonic fighter aircraft. A typical set of end-use requirements includes the axial or 0° compressive ultimate strength, modulus, and ultimate strain under room temperature dry conditions and the compressive ultimate strength, modulus, and ultimate strain under elevated temperature wet conditions. Generally, designers desire a specific strength, strain, and modulus at prescribed test conditions. Test conditions are selected to measure or gauge probable performance under in-service conditions. For example, for civil aircraft an often cited specification value is a minimum of 1034 MPa axial compressive strength at 82 °C in a wet condition. Military aircraft may require either a specific strength or strain at even higher temperatures.

Table 1 lists typical test results for standard epoxies and damage tolerant epoxies. Table 2 lists typical tests results for bismaleimide and thermoplastic based composites.

Tables 1 and 2 are very specific concerning the definition of modulus used. Uniaxial composites exhibit a non-linear stress-strain curve in tension and

Table 1. 0° Compressive properties for carbon fiber epoxy composites[a]

		Matrix	
		Standard epoxy	Damage Tolerant epoxy
Room temperature, dry			
	Strength, (MPA)	1725	1725
	Modulus[b], (GPa)	150	150
	Strain, (%)	1.1	1.1
Elevated temperature, wet			
	Strength, (MPa)	1450	1350
82 °C	Modulus, (GPa)	150	150
	Strain, (%)	1.0	0.9
	Strength, (MPa)	1210	1175
104 °C	Modulus, (GPa)	150	150
	Strain, (%)	0.8	0.8

[a] Fiber volume fraction = 62%, intermediate modulus fiber
[b] 0.1 to 0.6% Secant modulus

Table 2. 0° Compressive properties for carbon fiber BMI and thermoplastic composites[a]

		Matrix	
		BMI	Thermoplastic
Room temperature, dry			
	Strength, (MPa)	·1725	1070
	Modulus[b], (GPa)	150	150
	Strain, (%)	1.0	0.7
Elevated temperature, wet			
	Strength, (MPa)	970	–
163 °C	Modulus, (GPa)	150	–
	Strain, (%)	0.7	–

[a] Fiber volume fraction = 62%, intermediate modulus fiber
[b] 0.1 to 0.6% Secant modulus

compression. Hence, the strain range over which the modulus is measured has a strong effect on the value of that modulus. Generally, the higher the strain increment i.e. furthest removed from the origin (in tension) the higher will be the governing modulus. Details of composite stress-strain non-linearity will be addressed in Chap. 9.

Real life composite structure contains machined holes (for fastening) and/or can sustain through-hole damage. It is important that the structure be able to carry compressive (and tensile) loads in the presence of these stress risers. Hence, open-hole-compression (OHC) tests have emerged as a key measure of composite performance [29].

Computer modelling [30] of composites containing a hole and subjected to tension or compression have been developed for predicting both failure and strength of the composite. Information on failure modes and failure mechanisms with different ply orientation and load direction of CF/epoxy system was presented. This modelling technique may be helpful in defining composite strength and design of optimal composites structures with holes or cutouts.

8.4.1 Open-Hole-Compression Testing

Test laminates for open-hole-compression testing are generally quasi-isotropic i.e. $(+45/0/-45/90)ns$ and are commonly evaluated as a function of thickness or number of plies; e.g. 16, 24, 32, and 48 ply configurations are common. Specimen widths up to 125 mm with 25 mm diameter holes are tested under the NASA 1092 specification. Other end-users limit the test coupon width to

Table 3. Composite open-hole-compressive properties[a]

		Typical epoxy
Room temperature, dry		
	Strength, (MPa)	315
	Modulus[b], (GPa)	52
	Strain, (%)	0.6
Elevated temperature, wet		
	Strength, (MPa)	293
82 °C	Modulus, (GPa)	50
	Strain, (%)	0.6
	Strength	275
121 °C	Modulus, (GPa)	47
	Strain, (%)	0.6

[a] Lay-up $(0/\pm 45/90)_{3s}$, 6 mm hole, 200 × 50 mm coupon with 64 mm gauge length. Fiber volume fraction = 60%, Carbon fiber

[b] 0.1 to 0.6% Secant modulus

37.5 mm with a 6.25 mm diameter hole. The usual specimen length is from 250 to 312.5 mm.

Special care must be taken to ensure that the test specimen ends are ground flat and parallel. Test fixtures are used to introduce the compressive end load and support the specimen along the (long) edges to prevent a premature buckling failure. OHC test data provide guidance to designers who must ensure structural integrity under real life conditions which include cut-outs and holes and the environment. Tests are conducted at room and cold temperatures under dry conditions and under wet conditions at elevated temperature. Table 3 lists typical OHC test data for epoxy/carbon fiber composites; for a variety of test conditions.

8.5 Summary

Structure property guidelines governing uni-axial or longitudinal composite performance have been reviewed by emphasizing those constituent properties that require improvement for prescribed failure modes. Commonly used test techniques were defined and the importance of careful sample preparation and flawless test methodology was stressed. Testing in the presence of stress risers i.e. cut-outs, holes, damage, is required for real world composite structural applications. Open-hole-compression testing meets that need. Typical 0° compressive and OHC strength data has been presented for a number of different composite systems for both room temperature dry and elevated wet test conditions.

8.6 References

1. J. M. Alper and A. S. D. Wang, Comp. Matl's. Tech. Rev., 1981, 156.
2. M. R. Piggot, Journal Matl's Science, 16, 1981, 2837.
3. J. H. Sinclair and C. C. Chamis, NASA Technical Memorandum 82833, 1982.
4. H. T. Hahn, Tough Composites Workshop, NASA Langley Research Center, 1982, 72.
5. A. S. Wronski and T. V. Parry, Journal of Matl's Science, Vol. 17, 1982, 3656.
6. D. H. Woalstencroft, R. I. Harescuegh and A. R. Curtis, Progress in Science and Engineering of Composites, Tokyo, 1982, 439.
7. T. V. Parry and A. S. Wronski, Journal of Matl's Science, Vol. 17, 1982, 893.
8. B. Budiansky, Computers and Structures, Vol. 16, No. 1–4, 1983, 3.
9. R. Jones, W. Broughton, R. F. Monsley and R. T. Potter, Comp. Struct., 3, 1985, 167.
10. K. Tsukui et al., Proc. 17th National SAMPE Tech Conf., Oct. 1985, 40.
11. H. T. Hahn and J. G. Williams, Composite Materials: Testing and Design (Seventh Conf.), ASTM STP 893, ASTM, 1986, 115.
12. H. T. Hahn, M. Solu and S. Moon, NASA Contractor Report 3988, 1986.
13. E. Wilkinson, T. V. Parry and A. S. Wronski, Composites Science and Technology, 26, 1986, 17.

14. H. T. Hahn and M. Solu, Composites Science and Technology, 27, 1986, 25.
15. H. T. Hahn, Proc. of 6th ICCM and 2nd ECCM, Vol. 1, 1987.
16. R. J. Lee, Composites, Vol. 18, No. 1, Jan 1987, 35.
17. S. R. Allen, Journal of Materials Science, 22, 1987, 853.
18. A. J. Barker and V. Balasundaram, Composites, Vol. 18, No. 3, July 1987, 217.
19. S. W. Yugartis, Composites Science and Technology, 30, 1987, 279.
20. E. S. Zelenski et al., Proc. of 7th ICCM and 2nd ECCM, Vol. 1, 1987.
21. S. Kumar, W. W. Adams and T. E. Helminiak, Journal of Reinf. Plastics and Comp., Vol. 7, March 1988, 108.
22. S. J. DeTeresa, R. S. Porter and R. J. Ferris, Journal Matl's Science, 23, 1988, 1886.
23. D. Purslow, Composites, Vol. 19, No. 5, Sept 1988, 358.
24. P. S. Steif, Journal Comp. Matl's Science, 16, 1981, 2837.
24a. J. M. Prandy and H. T. Hahn, 35th International SAMPE April 1990, p. 1657.
24b. S. Kumar and T. E. Helminiak, SAMPE J. 26(2), 51 (1990).
25. C. C. Gedney et al., Proc. 32nd International SAMPE Symp., April 1987, 1015.
26. S. R. Swanson, G. E. Calom, Jr., and C. L. Halsom, Proc. 33rd International SAMPE Symp., March 1988, 1571.
27. T. A. Bogetti, J. W. Gillespie and R. B. Pipes, Composites Science and Technology, 32, 1988, 57.
28. D. F. Adams, 34th International SAMPE May 1989, 1422.
29. NASA Reference Publication 1092, 1983.
30. F-K Chang, L. Lessard, K-Y Chang and S. Liu, 34th International SAMPE May 1989, 559.

9 Damage Tolerant Composites: Post Impact Compressive Strength

9.1 Introduction

The need for toughened or damage tolerant composites arose in the late 1970s when airlines were faced with escalating fuel costs and hence the need for lighter-weight, more fuel-efficient aircraft. Airframe companies identified toughened or damage tolerant composites as a means to reduce aircraft weight. A key requirement of damage tolerant composite structures was the ability to sustain compressive loading in the presence of impact induced damage. Critical structural components included wing skins which could involve thick (6.25 mm or greater) laminates. A real life complication arising from thick laminates is that impact damage is generally not visible from the impact side i.e. the impact event often causes a delamination zone in the laminate interior and/or reverse side damage, neither of which are readily assessed.

Additional key requirements for achieving structural lightweighting focussed on carbon fiber reinforcement. Specifically, it became evident that carbon fibers with a true 2% ultimate strain were necessary. Damage tolerant composite structure also benefited from higher stiffness lamina. This motivated the development of intermediate modulus carbon fibers, i.e. a fiber axial modulus of 276–289 GPa versus the more traditional carbon fiber modulus of 227–241 GPa.

Designers developed criteria to define acceptable composite damage tolerance based on laminate compressive strength-after-impact or post-impact-compressive strength. Another key performance requirement was to maintain an absolute axial or longitudinal composite-compressive-strength in the presence of environments at low and/or elevated temperatures. Hot/wet compressive testing was discussed in Chapter 8.

Damage tolerance is also important for thin structure (often in the range of 0.75 to 1.5 mm) whereby an impact event results in through laminate penetration i.e. a hole. In this case the structure must sustain compressive (or tensile) loads in the presence of the stress riser (hole) and/or harsh environment. Open-hole-compressive strength was also discussed in Chap. 8.

Air frame companies, NASA, and the materials suppliers industry actively studied damage tolerant composites throughout the 1980s. Many new composite systems were developed and testing methodology has evolved for characterizing damage tolerance in composites. Composite toughness as it relates to post-impact residual compressive strength will be discussed with a focus in the emerging understanding of structure property relationships applied to the developing area of damage tolerance. Data governing characteristics of current "tough" thermoset and thermoplastic matrix/carbon fiber reinforced composites will be presented.

9.2 Compressive-Strength-After-Impact

Early experimental investigations in composites toughness have been reported extensively by many investigators [1–6] who evaluated composites based on both woven and uni-directional lamina. Static compressive and cyclic compressive behavior was characterized for laminates with circular holes, simulated delaminations, and after low-velocity impact. Generally, the laminates investigated were quasi-isotropic with dimensions of 100 mm by 150 mm by 5 mm.

The use of the width-tapered double-cantilever beam specimen for assessing composite toughness was advanced. Conclusions reached indicated that the most damage tolerant composites exhibited the least delamination within the cross-section due to impact, had the highest transverse ultimate tensile strain, and the largest crack opening force as determined from double-cantilever-beam tests.

Another important result was that the addition of elastomeric modifiers to the matrix resin led to nearly a power of ten increase in fracture energy and hence apparent composite toughness.

Cooperative efforts between the aircraft industry and NASA led to the publication of standardized tests for toughened resin composites in 1983 [7]. Additional joint NASA/aircraft industry efforts led to the refinement and expansion of the criteria of NASA 1092, and in 1985, NASA Reference Publication 1142 defined standard specifications for graphite fiber/toughened thermoset resin composites [8].

9.2.1 Standard Tests

Table 1 lists the standard tests for toughened resin composites which emerged from the joint NASA/aircraft industry efforts. Advances were also made in the test methodology and analysis for the double-cantilever beam test, edge delamination test, and various drop weight testing [9–15].

Two laminate configurations emerged for defining compressive strength-after-impact. The specimen described in NASA 1092 has a ply orientation of $[+45/0/-45/90]_{ns}$ and a nominal thickness of 6.25 mm. The specimen size is 175 by 375 mm before impact and 125 × 312.5 mm after impact. Test fixtures have been defined to provide lateral support against buckling and a means of introducing the end-load without end-crushing. The second commonly used laminate is also quasi-isotropic and is nominally 100 × 150 × 6.25 mm.

Compression-after-impact testing requires extreme care in specimen preparation; edges must be parallel and true. Fixtures are used during the sample

Table 1. Standard tests for toughened resin composites (NASA 1092)

Test	Ply orientation	Nominal thickness (mm)	Test specimen size (mm)
Compression after impact	$[45/0/-45/90]_{ns}$	6.25	175 by 375 (before impact) 125 by 312.5 (after impact)
Edge delamination tension	$[\pm 30/\pm 30/90/90]_s$ $[\pm 35/0/90]_s$	11 plies 8 plies	100 by 150 by 6.25
Open-hole tension	$[45/0/-45/90]_{ns}$	6.25	50 by 300
Open-hole compression	$[45/0/-45/90]_{ns}$	6.25	125 by 317.5
Double cantilever beam	$[0]_n$	1.65 ± 0.18	12.5 by 200

impact. The stiffness/energy absorbing characteristics of this support fixture must be controlled to ensure that a uniform amount of energy from impact is absorbed by the specimen.

In general, the damage tolerance of a composite is judged by the level of residual compressive strength after a defined impact.

Impact conditions are defined in terms of a specific energy level for a given thickness of laminate, J/m. A commonly used impact value is 6700 J/m with a target residual compressive strength of 310–345 MPa.

9.3 Toughened Matrix Resins

A general discussion of epoxies, bismaleimides, and toughened bismaleimides was presented in Chapter 2. The development of toughened thermosetting matrices during the 1980s, with special emphasis on neat resin properties, has been effective in identifying resin systems with increased damage tolerance in composites. Earlier efforts dealt with epoxy systems designed for hot/wet performance in the 82° to 104 °C. More recently, toughened BMI formulations designed for hot/wet service to 177 °C have been developed.

In the development of toughened thermosets, a balance of properties has been sought. Specifically, formulations must be prepregable for proper degree of tack and drape, have adequate modulus and T_g, ideally possess low equilibrium water sorption, and have increased toughness relative to baseline matrices. In general, modifications to matrices, morphological changes or introduction of additives, all of which result in dramatic increases in toughness, can also lead to increased water sorption. Modifications of formulations which have high glass transition temperatures (T_gs) and low water sorption are often quite brittle. Hence, the achievement of a balanced neat resin property profile is a formidable challenge.

Moulton and Ting [16] conducted some of the earliest work to define the effects of elastomeric additives on composite toughness. They studied acrylonitrile-butadiene (CTBN) modified epoxies reinforced with both E-glass and carbon fibers (Thornel T300). Elastomeric modifiers greatly increased the toughness of the neat resin as measured by compact tension and Izod impact. The extent of increased toughness was dependent on the amount and molecular weight of the CTBN additives. However, matrix modulus, strength, water sorption, and use temperature were compromised.

NASA sponsored research [17, 18] investigated twenty-four epoxy matrix resins, reinforced with Thornel T300 carbon fiber. Damage tolerance was evaluated based on impact induced damage in a 48 ply orthotropic laminate with ply orientation $[\pm 45/0_2/\pm 45/0_2/\pm 45/0/90]_{2s}$. In addition, unidirectional laminate shear and compressive properties were measured. The five materials which exhibited the greatest damage tolerance contained resin modifiers consisting of elastomers, thermoplastics and vinyls. In all cases the base epoxy was bisphenol A. Data indicate that to achieve damage tolerance under low velocity impact conditions, the room-temperature neat-resin tensile-property profile should include: a tensile strength greater than 69 MPa, a tensile modulus of at least 3100 MPa, and an ultimate tensile strain between 4 and 5%. Low longitudinal composite strength was observed for neat resins with a tensile modulus below 3100 MPa.

The investigation of damage tolerant epoxies for structures other than bisphenol A has been carried out. Results reported include data on neat-resin fracture toughness as well as composite damage sustained under low velocity impact conditions [19–36].

An excellent summary of the work on toughened composites is a recent publication by Johnson [28]. This work contains the results of both government and industry sponsored research and includes composites based on both thermoplastic and thermosetting matrices.

9.4 Inter Leaf Concept

An unusual technique for improving impact resistance is through the use of an interleaf concept. It consists of interleafing a film adhesive containing a thermosetting or thermoplastic resin between plies of a laminate. Krieger [32] reported the use of an adhesive interleaf between plies of a laminate as an effective means of achieving enhanced toughness. The interleaf material exhibited high strain and low-flow and was effective in reducing stress concentrations as well as inhibiting growth of impact induced damage. The properties trade-off included elevated temperature wet performance and longitudinal compressive strength. Masters [33] has described a tough, ductile thermoplastic resin film as the interleaf component for a toughened, epoxy system. Impact resistance improved with a minimal reduction in structural performance.

9.5 3-D Reinforced Fiber Preforms

A novel way to make more damage tolerate composites is by designing 3-D reinforced fiber preforms [34] via new fiber technologies with minimal fiber damage. Tests have shown that the 3-D reinforcement results in better interlaminar properties, better damage tolerance and improved fracture mechanics. The technique involves the use of resin transfer molding (RTM) to fabricate the improved damage tolerant composite part. The rationale is to use "textile preform architecture" by stitching, knitting, braiding, etc., (near net shape preform) which introduces the improved damage tolerance and then treat the preform with a thermosetting resin. Toughness in the fabricated part is provided through "stitching" versus the presently accepted technique of using toughened resins (modified thermosets or thermoplastics). A further refinement of the 3-D method consists of 3-D braided thermoplastic commingled (see Thermoplastic Composites) preform [35]. The effects of braid architecture and degree of fiber wet-out translated into improved open hole tension, open hole compression and compressive strength after impact.

9.6 Structure Property Trends

Studies to define the desired characteristics of carbon fiber to achieve optimal composite performance, including toughness, are reported by McMahan [37] and Farley [38]. Impact performance of composites based on both thermoplastic and epoxy matrices is discussed by Dorey and coworkers [39]. Additional progress [40–44] in understanding damage tolerance of composite laminates includes descriptions of test technique, post-impact shear and fatigue, damage zone growth, and the effect of fiber type. Manders and Harris [45] published the results of a systematic study of compression-after-impact performance which showed that fiber/matrix adhesion, matrix toughness, and fiber volume fraction are the three variables which have the greatest effect on damage tolerance. The parametric study of compression-after-impact included the following variables: carbon fiber type (strength, modulus, surface functionality); matrix type (toughened, baseline, resin gelation); fiber volume fraction (52 to 60%); ply thickness and lay-up (145 g/m^2 vs 190 g/m^2, quasi-isotropic 24 ply vs 36 ply anisotropic); and test orientation (0° vs 90° plies in the center of the lay-up). Key conclusions attained include: an optimal degree of fiber surface functionality exists to promote the proper degree of adhesion between the matrix and the fiber, too low adhesion adversely affects damage tolerance; fiber tensile strength has little influence on damage tolerance within the range of 3100–5500 MPa; for a given fiber, matrices with higher ultimate tensile strength result in improved CAI values; higher resin content laminates (52 to 60% fiber volume) give higher CAI values; the commonly used variations on lay-up, ply thickness, and orientation of the quasi-isotropic laminate in the compression test, have no significant effect on compression-after-impact results.

Recently Stuart and Altstadt [46] have reported some studies which deal with characterization and effect of resin-fiber interface as it applies to composite damage tolerance performance of modified BMI/carbon fiber systems. With higher levels of surface treatment, it was noted that fracture initiation predominately occurred in the resin matrix rather than at the fiber-resin matrix interface. More importantly, they observed a 20% increase in CAI with increasing surface treatment.

9.7 Representative Damage Tolerant Systems

Damage tolerant composite development during the past five years has incorporated the concepts of toughened matrices, prepreg morphology, fiber characteristics, and test techniques discussed in the preceding sections. Table 2 summarizes the performance of current damage tolerant systems as of 1990 [47–51] and the companies that supply the systems. Bismaleimide based composites which highlight improved resistance to microcracking and enhanced damage tolerance for composites suitable for wet service to 177 °C are described [52–55].

Table 2. New damage tolerant epoxy prepregs

Supplier	System[a]	CAI strength[c] (MPa)	Wet use temp (°C)
American Cyanimid	1840/IM7	310	82
Amoco Performance	1982/T650-42	310 to 345	82 to 104
Products	1983/T650-42	276	≥ 104
Ciba Geigy	6377/IM7	345	≥ 82
Fiberite[b]	977-2/IM7	262 to 303	104
Hercules	8551-7A/IM7	310 to 345	104
Narmco	5255-3/IM7	310 to 345	82
Toray	3900-2/T800	345	82
US Polymeric	E7T1/IM7	310	≥ 82

[a] Matrix resin/Fiber type.
[b] All systems cured at 177 °C for 2 h except 977-2 which requires 3 h at 177 °C for full cure.
[c] Nominal fiber volume 58% and 6700 J/m impact level.

9.8 Summary

Research and product development efforts during the 1980s resulted in a number of epoxy/intermediate modulus carbon fiber systems with compression-after-impact performance in the 310–345 MPa range with wet service temperatures from 82 to 104 °C. This represents an improvement over existing baseline systems which were generally suitable for wet service to 104 °C but had CAI values in the vicinity of 138–165 MPa. In addition, higher wet service systems (177 °C) based on bismaleimides have also shown significant improvement in toughness. A current area of high interest is processability of composite parts with damage tolerant prepregs. Specifically, optimizing prepreg tack, drape, release, and out-time to permit fabrication via automatic tape laydown (ATL) is a major thrust. Additional efforts focus on the fabrication of thick structures (90 or more plies thick) free of voids or other (gaps) lay-down/cure induced flaws.

9.9 References

1. M. D. Rhodes, NASA TM78755, 1978.
2. M. D. Rhodes et al., 34th Ann. Tech. Conf. Reinforced Plastics/Comp. Instrt., SPI, 1979.
3. A. G. Miller, P. E. Hertzberg and V. W. Rantala, National SAMPE Tech. Conf. 12th 1980, 279.
4. B. A. Byers, NASA Contractor Report 159293, August, 1980.
5. W. D. Bascom, J. L. Bitner, R. J. Moulton and A. R. Siebert, Composites, Vol. 11, No. 1, 1980.
6. A. K. Green and W. H. Bowyer, Composites Vol. 11, No. 7, 1980, 131.
7. NASA Reference Publication 1092, May 1983.
8. NASA Reference Publication 1142, 1985.
9. J. M. Whitney et al., J. Reinforced Plastics and Composites, Vol. 1, Oct. 1982, 297.
10. T. K. O'Brien et al., SAMPE Journal, July/Aug. 1982, 8.
11. N. J. Johnston et al., 28th National SAMPE Symp. & Exh. April 1983.
12. M. W. Wardle and E. W. Tokarsky, Comp. Tech. Rev., Vol. 5, No. 1, 1983, 4.
13. J. D. Winkel and D. F. Adams, Composites, Vol. 16, No. 4, 1985, 268.
14. D. A. Scola et al., SAMPE Journal, March/April 1986, 47.
15. S. A. Thompson and R. J. Farris, SAMPE Journal, Jan./Feb. 1988, 47.
16. R. J. Moulton and R. Y. Ting, Compos. Struct. [Proc. 1st Int. Conf.], 1981, 674.
17. R. J. Palmer, NASA Contractor Report 165677, March 1981.
18. J. G. Williams and M. D. Rhodes, NASA Technical Memorandum 83213, October 1981.
19. W. D. Bascom and D. L. Hunston, Adhesion, Vol. 6, 1982, 185.
20. P. E. Hertzberg et al., NASA Contractor Report 165784, Jan., 1982.
21. A. F. Yee and R. A. Pearson, NASA Contractor Report 3718, Aug., 1983.
22. G. D. M. DiSalvo and S. M. Lee, SAMPE Quarterly, Jan. 1983, 14.
23. M. Ashizawa, Proc. 6th Conf. on Fibrous Comp. in Struc. Design, Jan. 1983.
24. J. G. Davis, Jr., (Ed), NASA Conference Publication 2321, Aug. 1984.
25. Anom., NASA Contractor Report 172 358, Aug. 1984.
26. J. Diamant and R. J. Moulton, SAMPE Quarterly, Oct. 1984.

27. J. M. Morgahs, Plastics Design Forum, Sept./Oct. 1985, 78.
28. N. J. Johnston, (Ed), Toughened Composites ASTM Special Technical Pub. 937, 1987.
29. K. J. Bowles, NASA Technical Memorandum 87337, Apr. 1986.
30. I. Gawin, Proc. 31st Inter. SAMPE Symp., April 1986, 1204.
31. S. M. Arndt, Annual Forum Proc., American Helicopter Society, Vol. 1, 1991, 653.
32. R. B. Krieger, Jr., SAMPE Journ., July/Aug. 1987, 30.
33. J. E. Masters, et. al., SAMPE 31, 844 (1986); SAMPE 34, 1792 (1989).
34. F. J. Arendts, K. Drechsler and J. Brandt, 34th International SAMPE, May 1989, 2118.
35. C. T. Hua and F. K. Ko, 21st International SAMPE Tech. Conf. Sept. 1989, 688.
36. J. Y. Liau et al., Plastics Engineering, Nov. 1988, 33.
37. P. E. McMahon and D. F. Taggart, Progress in Science and Engineering of Composites, T. Hoyoski, K. Kawata, and Umekawa, eds., ICCMIV, Tokyo, 1982, 529.
38. G. L. Farley, Journal of Composite Materials, Vol. 20, July 1986, 322.
39. G. Dorey, S. M. Bishop and P. T. Curtis, Comps. Sci. and Tech., 23, 1985, 221.
40. K. Kanumuira et al., 4th International SAMPE Conf., Bordeaux, France, 1983, 127.
41. S. M. Lee, SAMPE Journal, March/April, 1986, 64.
42. K. D. Challenger, Composite Structures, 6, 1986, 295.
43. D. J. Ball et al., Journal of Materials Science, 21, 1986, 2667.
44. S. Lloreute, 43rd Annual National Forum American Helicopter Society, May 1987, 43.
45. P. W. Manders and W. C. Harris, SAMPE Journal, Nov./Dec., 1986, 47.
46. M. Stuart and V. Altstadt, 21 International SAMPE Tech. Conf. Sept. 1989, 264.
47. G. R. Almen et al., 34th International SAMPE Symp., May 1989, 259.
48. F. K. Chang et al., 34th International SAMPE Symp., May 1989, 702.
49. L. D. Bravenec et al., 34th International SAMPE Symp., May 1989, 714.
50. H. G. Recker, 34th International SAMPE Symp., May 1989, 747.
51. R. S. Bauer et al., 34th International SAMPE Symp., May 1989, 899.
52. A. Bosch, 21st International SAMPE Tech. Conf. Sept. 1989, 275.
53. M. M. Konarski, 34th International SAMPE Symp., May 1989, 514.
54. V. Ho, 34th International SAMPE Symp., May 1989, 514.
55. J. D. Boys et al., 34th International SAMPE Symp., May 1989, 2365.

10 Thermoplastic Composites

10.1 Introduction

The potential of emerging, carbon-fiber-reinforced thermoplastic-matrix composites for structural applications versus conventional thermosetting polymeric-matrix-based composites was recently reviewed [1]. Summaries [2, 3, 4] of current developments suggest that thermoplastic composites afford the potential for elevated temperature/wet service approaching 210 °C, enhanced durability (toughness and repairability), and automated, cost-effective processing techniques. These factors could result in favorable life cycle economics based on materials acquisition cost, performance, producibility and supportability.

Certain thermoplastic composites like polysulfone-based systems [3] were studied extensively during the 1970s. Mechanical property data bases and processing technologies were developed through the prototype hardware stage to demonstrate cost and performance advantages. Thermoplastic composites technology has evolved and is continuing to evolve rapidly. The data base and manufacturing experience needed for wide spread industrial application is currently being established.

Carbon fiber reinforced thermoplastic composites will be the primary focus for review. There are other thermoplastic resin reinforcing fiber systems (aramid, S-glass) but the major thrust of thermoplastic composites has been directed to carbon fiber for aircraft structural applications. The focus will be on promising high temperature matrices, product forms, processing or fabrication technologies, typical composite properties and issues being addressed in current and developmental programs.

10.2 High Performance Thermoplastic Matrices

The amorphous and crystalline high performance thermoplastics that gained acceptance as composite matrix resins were discussed in Chapter 2. Neat resin, thermal, physical, and mechanical properties important for composite processing and performance have been presented in Table 9 of Chap 2. Table 1 summarizes the glass transition temperatures (T_gs) and processing temperatures for a number of matrix resins. More complete information concerning polymer trade name and manufacturer can be obtained by examining Ref. 2. Additional neat resin data for polyaryl ethers are provided in Ref. 3. Polyamide-imide

Table 1. Thermoplastic matrix thermal characteristics

	Glass transition, T_g (°C)	Processing temperature (°C)
Polymer		
Polyamide-imide (PAI), (A)[a]	275	345–355
Polyaryl ethers (A)	220–260	310–345
Polyether sulphone (PES), (A)	220	300–310
Polyether-imide (PEI), (A)	210–200	345
Polyarylene sulfide (PAS), (A)	200–210	330–345
Polyetherether ketone (PEEK), (C)	140–145	340–350
Polyphenylene sulfide (PPS), (C)	85–95	330
Polyarylene ketone (C)	200–210	370–415
Polyimide (PI), (A, C)[a]	250–280	350–360

(C) Crystalline, (A) Amorphous
[a] Polymers sometimes characterized as pseudothermo-plastic based on limited reformability.

properties are summarized in Refs. 5 and 6 while polyimide properties are available in Refs. 7–10.

Liquid crystalline polymers such as Xydar and Vectra were also discussed in Chap. 2. These polymers afford the combination of high glass transition temperatures (350 °C) and high mechanical properties which make them viable candidates for very high performance composite applications. However, composite studies with liquid crystalline polymers are at a very early stage.

Composites based on the crystalline polymer, poly ether ether ketone (PEEK) have been extensively investigated and have resulted in structural aerospace applications over the past decade. Polyarylethers (Amoco's Radel 8320) are amorphous polymers and have been thoroughly studied in emerging aircraft applications within the last few years. Composite data bases for both PEEK and Radel matrices are becoming well established for a number of different carbon fibers including intermediate modulus PAN fibers as well as high modulus (P-75) and ultra high modulus (P-100) pitch based fibers. Some special concern for composites based on crystalline matrices is expressed by Vautey [41]. Loss of toughness due to an increase in crystallinity was observed during CAI testing. Controlled cooling rates are necessary for the desired degree of crystallinity or morphology.

10.3 Thermoplastic Prepreg Product Forms

Pre-impregnated fiber, more generally designated, "prepreg" represents an intermediate material form for fabrication of composite laminates or components. Important characteristics of prepregs include: degree of fiber impregnation, resin or fiber content, tack, drape, flow, storage stability and outlife or tool life. In general thermoplastic-matrix-based, continuous-filament prepregs achieve a consistent degree of fiber wet out, uniform fiber distribution, no tack, no drape, i.e., boardiness, adequate flow under proper process conditions and unlimited storage/tool life. Molyneux [11] and Seferis [12] have described the basics of prepreg processes considering both thermosetting and thermoplastic matrix resins. Solvent or solution technologies as well as hot melt techniques are discussed. Special processing steps are required to accommodate the high viscosity of thermoplastics. Melt viscosities of 10^4 to 10^6 Pa s at temperatures above 200 °C are common for thermoplastics as compared to 10^3 to 10^4 Pa s for thermosets at room temperature.

Hot-melt processing in the absence of tackifiers results in a stiff/boardy prepreg for both unidirectional tape and woven fabrics. The advantage is the lack of residual solvent. Solvent or solution impregnation can result in either boardy, tack-free prepreg or prepreg with drape and tack similar to a conventional thermosetting system when high levels of solvents are employed. The disadvantage of high residual solvent thermoplastic prepreg is two fold: limited storage/tool life and process design for solvent removal/management during part fabrication or consolidation. Solvent removal may result in void/blister formation.

In the case of thermoplastic matrix resins, a number of prepreg forms have emerged to meet the requirements of various fabrication processes.

10.3.1 Quadrax

Unidirectional prepreg tape, generally available in 30-cm width, can be slit into narrow tapes, typically 3.2 mm, to approximate tow prepreg. Slit tapes or tow prepreg can be woven [13] into various style fabrics. This results in a "preform" with more or less drape depending on the particular weave style employed. The degree of drape achieved enhances the ability to conform to tooling contours.

10.3.2 Electrostatic/Powder

New technologies are being commercialized [14, 15] for the production of either prepreg tow or tape prepregs via electrostatic or other powder impregnation methods. In some cases special additives and processing steps are reported to yield thermoplastic prepreg with sufficient tack to approximate thermoset lay-up.

10.3.3 Commingled Yarns

Hybrid or commingled yarns of carbon fibers and thermoplastic filaments [16, 17, 18] have emerged as an important product form. The commingled yarns

Table 2. Thermoplastic prepreg forms

	Tack	Drape	Width (mm)
Unidirectional Tow and Tape			
Solvent process	possible	possible	3–300
Hot-melt process	no	no	
Powder process	possible	no	
Woven Fabric			
Solvent or hot-melt	no	no	> 600
Commingled Woven Fabric	no	yes	> 600
Woven Slit Tape	no	some	3000

can be processed as tows or woven into fabric. The weave style of the fabric can be selected to optimize drape for consolidation into specific parts. Critical to achieving well-consolidated, void-free composites from commingled-woven fabric is adequate polymer flow to accomplish impregnation of the fiber bundles. Fiber wet-out is enhanced by matching the diameters of the thermoplastic filament and the carbon fiber and by intimate contact of filament and fiber. Programs to achieve thermoplastic fine filament diameters and optimized commingling continue in earnest. Fabric cowoven from carbon fiber and thermoplastic filaments which were not commingled result in poor quality consolidated parts due to inadequate polymer flow for the filament diameter and degree of interfilament contact. Programs are also addressing the use of commingled tow in woven interlocked preforms to achieve enhanced out of plane property performance [19]. Table 2 lists thermoplastic composite prepreg forms.

10.4 Processing Technology

Processing thermoplastic composites has become increasingly sophisticated since early efforts [3] that focused on stacking alternate plies of woven reinforcement and thermoplastic film followed by press consolidation under an appropriate pressure–temperature cycle. Most standard thermosetting composite processing techniques are applied to thermoplastics with special adaptation of metal forming techniques in recognition of the boardy, tack-free nature of thermoplastics [20, 21]. Since no chemical 'crosslinking' reaction or cure is needed for thermoplastics, a four or five fold reduction in process time (vs thermosets) is anticipated.

Technology based guidelines [22] for the processing of thermosets have been highly developed. Similar guidelines on the material response of thermoplastic composites during processing is now emerging [23–25].

Table 3. Thermoplastic processing methods

Method	Reference
Autoclave consolidation	30
Press forming (rubber assisted punch or hydro forming)	30, 31
Double diaphragm forming	30, 31
Filament and tape winding	32, 33, 34
Pultrusion	32, 35, 36, 42
Roll forming	30

Thermoplastic prepreg, prior to consolidation and forming, must, like thermosets, be arranged into laminates with the desired fiber orientation in each ply. The process for creating the desired oriented ply stack-ups has developed from manual spot and seam fusion welding of prepreg to the use of automated tape lay-up machinery [26]. The manufacturing efficiencies afforded by automation hold promise for cost effective production of thermoplastic parts. Details of machinery and thermal mechanical modelling have been published [27–29] for specific materials. The results are applicable to most thermoplastics.

A recently published paper [30] describes in detail the fabrication of a thermoplastic-composite, fighter-forward-fuselage demonstration article. Manufacture of this part demonstrates a wide range of tooling, fabrication, and assembly methods for thermoplastic composites. Five specific material systems were evaluated including amorphous and crystalline resins, and unitape, unifabric, and woven fabric product forms. Results of this study indicate that autoclave consolidation is the best method for producing large parts. Double diaphragm forming allows the forming and consolidation of parts in one operation and has the potential for cost-effective production. Summaries of diaphragm processing for both metallic and polymeric diaphragms as well as press forming techniques using a rubber punch tooling concept suggest [30, 31] that these methods are viable for thermoplastic composite processing.

The techniques discussed above and other processing technologies important for thermoplastic processing are listed in Table 3. Two areas of extreme importance in any technology are the manufacturing costs and quality assurance methodology. Recent publications [43, 44] assess costs and QA respectively with pertinent details applicable to thermoplastic composites.

10.5 Assembly Techniques

The highly successful study [30] that detailed the fabrication of a thermoplastic composite fighter-forward-fuselage indicated that thermoplastic composite components can successfully eliminate a high percentage (73%) of mechanical fasteners since thermoplastic systems are amenable to adhesive, dual polymer bonding and consolidation to join formed details into component assemblies.

Dual polymer bonding is particularly effective with amorphous polymers like Amoco's Radel polyaryl ethers since thermodynamically miscible matrix resins (see Blends) are available with a difference in T_g of 30 to 40 °C. The use of lower T_g Radel film between details of higher T_g Radel based composites leads to strong, durable joints in a single consolidation step. Fusion bonding techniques applicable to joining thermoplastic composites have been summarized recently [37–40].

10.6 Polyarylether Sulfone/Carbon Fiber Composites

The amorphous thermoplastic poly(aryl ether sulfone) has proven to be an excellent composite matrix resin. Specifically, Amoco Radel reinforced with Thornel carbon fibers provides a readily processable solvent free composite system with a wet service temperature of 163 °C.

Radel 8320 based composites are available in several product forms including: uni-tape, commingled fabric, and Quadrax interlaced tape. This variety of prepreg product forms affords the opportunity to select the prepreg which has the combination of drapeability, process parameters and composite property profile which best suits the end-use application. In addition to several prepreg product forms, Radel 8320 prepreg is available with a range of carbon fiber reinforcement including standard modulus Thornel T650-35, intermediate modulus Thornel T650-42 and higher intermediate modulus Thornel T750-45 and Magnimite IM-8.

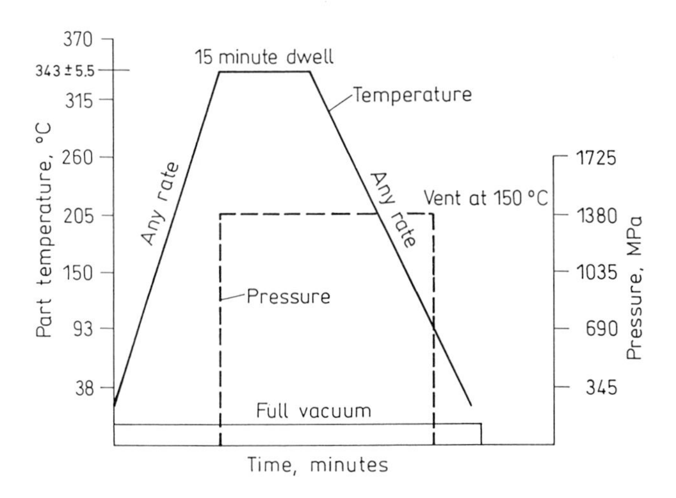

Fig. 1. Radel 8320 unitape consolidation cycle

10.6.1 Processing

Radel based composites have been successfully processed by all the processes mentioned in Table 3. However, the best developed data base has been generated for consolidation in an autoclave. The process parameters for autoclave consolidation of Radel 8320 composites are independent of carbon fiber reinforcement type but do depend on the prepreg product form. Figures 1 and 2 show that the consolidation cycle for unitape and Quadrax interlaced tape are identical.

The cycle involves a 15 minute dwell at nominally 343 °C and 1.39 MPa. Figure 3 illustrates consolidation of composites from Radel 8320 commingled

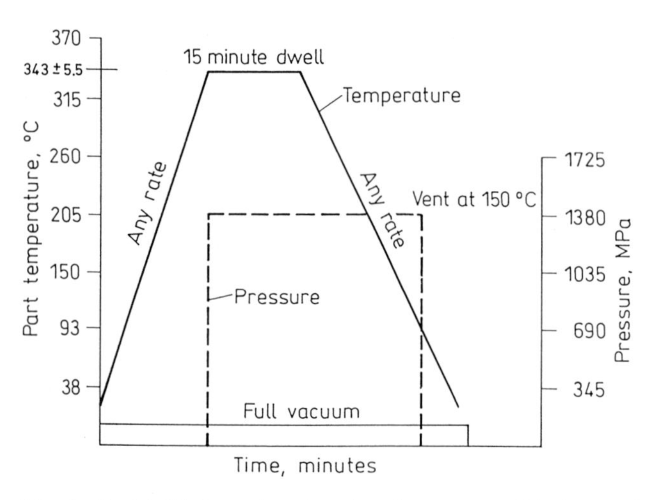

Fig. 2. Radel 8320 quadrax interlaced tape consolidation cycle

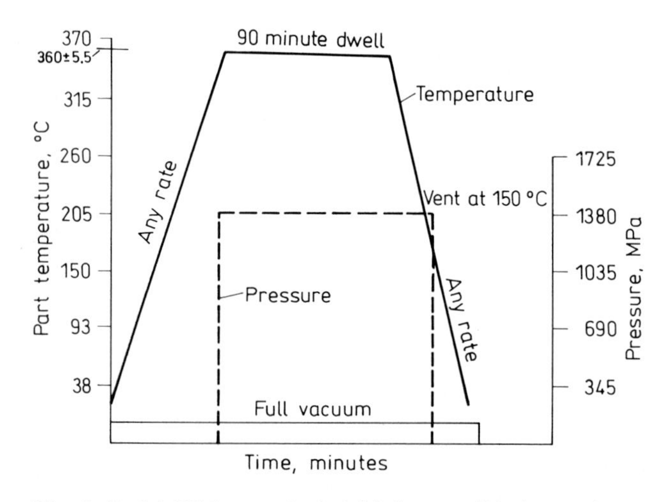

Fig. 3. Radel 8320 commingled fabric consolidation cycle

Table 4. Thornel T650-35/Radel 8320 composite properties

	Provisional Data Sheet (Vf = .62)			
Property	Unit	Value	Condition	Comment
0° tension (ASTM D-3039)				
strength	MPa	2158	RTD	
strain	%	1.39	RTD	
modulus	GPa	153	RTD	(1)
0° compression (ASTM D-695)				
strength	MPa	1027	RTD	
modulus	GPa	148	RTD	(2)
90° tension (ASTM D-3039)				
strength	MPa	77	RTD	
strain	%	1.02	RTD	
modulus	GPa	8	RTD	(1)
± 45 deg tension (ASTM D-3518)				
strength	MPa	242	RTD	
modulus	GPa	18	RTD	(2)
shear modulus	GPa	5	RTD	(3)
Edge delamination (NASA 1092)				
strength	MPa	352	RTD	
Open-hole compression (24 ply QI, 3.18 cm wide, 0.64 cm hole)				
strength	MPa	292	RTD	
strength	MPa	228	150 °C wet	(4)
Open-hole compression (24 ply QI, 3.81 cm wide, 0.64 cm hole)				
strength	MPa	305	RTD	
Open-hole tension (24 ply QI, 3.18 cm wide, 0.64 cm hole)				
strength	MPa	413	RTD	
Quasi-Isotropic tension strength (24 ply QI, 3.18 cm dogbone)				
strength	MPa	900	RTD	
modulus	GPa	56	RTD	(1)
Quasi-Isotropic compression strength (24 ply QI, 3.18 cm dogbone)				
strength	MPa	560	RTD	
modulus	GPa	54	RTD	(2)
Compression-after-impact (32 ply QI, 10.2 × 15.2 cm, 6673 J/m)				
strength	MPa	288	RTD	

(1) Secant from 0.1% to 0.6% strain
(2) Calculated at the origin
(3) Secant from 0.01% to 0.1% strain
(4) Conditioned @ 77 ± 3 °C in 80%–85% R.H. for 20 ± 1 days

Table 5. Thornel T650-42/Radel 8320 composite properties

	Provisional Data Sheet (Vf = .62)			
Property	Unit	Value	Condition	Comment
0° tension (ASTM D-3039)				
strength	MPa	2420	− 54 °C	
		2289	RTD	
strain	%	1.49	− 54 °C	
		1.38	RTD	
modulus	GPa	163	− 54 °C	(1)
		166	RTD	(1)
0° compression (ASTM D-695)				
strength	MPa	1069	RTD	
modulus	GPa	157	RTD	(2)
90° tension (ASTM D-3039)				
strength	MPa	61	RTD	
strain	%	0.79	RTD	
modulus	GPa	8	RTD	(1)
± 45 deg tension (ASTM D-3518)				
strength	MPa	274	RTD	
modulus	GPa	18	RTD	(2)
shear modulus	GPa	4	RTD	(3)
Edge delamination (NASA 1092)				
strength	MPa	424	RTD	
Open-hole compression (24 ply QI, 3.18 cm wide, 0.64 cm hole)				
strength	MPa	285	RTD	
Open-hole compression (16 ply QI, 3.81 cm wide, 0.64 cm hole)				
strength	MPa	269	RTD	
strength	MPa	241	105 °C wet	(4)
strength	MPa	225	135 °C wet	(4)
strength	MPa	208	150 °C wet	(4)
strength	MPa	192	163 °C wet	(4)
Open-hole tension (16 ply QI, 3.81 cm wide, 0.64 cm hole)				
strength	MPa	475	RTD	
Compression-after-impact (32 ply QI, 10.2 × 15.2 cm, 6673 J/m)				
strength	MPa	296	RTD	

(1) Secant from 0.1% to 0.6% strain
(2) Calculated at the origin
(3) Secant from 0.01% to 0.1% strain
(4) Conditioned @ 77 ± 3 °C in 80%–85% R.H. for 20 ± 1 days

fabric requires no additional pressure but an increase in the dwell temperature to 360 °C and an increase in the dwell time to 90 minutes. The added time at a higher temperature provides for heat transfer and resulting flow of the matrix to result in well impregnated carbon fiber required by the distinct morphology of commingled products.

Table 6. Quadrax format Thornel T650-42/Radel 8320 composite properties

	Provisional Data Sheet (Vf = .62)			
Property	Unit	Value	Condition	Comment
0° tension (ASTM D-3039)				
strength	MPa	1131	RTD	
strain	%	1.33	RTD	
modulus	GPa	84	RTD	(1)
0° compression (ASTM D-695)				
strength	MPa	585	RTD	
modulus	GPa	84	RTD	(2)
90° tension (ASTM D-3039)				
strength	MPa	–	RTD	
strain	%	–	RTD	
modulus	GPa	–	RTD	(1)
± 45 deg tension (ASTM D-3518)				
strength	MPa	199	RTD	
modulus	GPa	17	RTD	(2)
shear modulus	GPa	4	RTD	(3)
Edge delamination (NASA 1092)				
strength	MPa	–	RTD	
Open-hole compression (8 ply QI, 3.18 cm wide, 0.64 cm hole)				
strength	MPa	268	RTD	
strength	MPa	195	150 °C wet	(4)
Open-hole compression (8 ply QI, 3.81 cm wide, 0.64 cm hole)				
strength	MPa	–	RTD	
Open-hole tension (8 ply QI, 3.81 cm wide, 0.64 cm hole)				
strength	MPa	460	RTD	
Quasi-Isotropic compression strength (12 ply QI, 3.18 cm dogbone)				
strength	MPa	–	RTD	
modulus	GPa	–	RTD	
Compression-after-Impact (16 ply QI, 10.2 × 15.2 cm, 6673 J/m)				
strength	MPa	288	RTD	

(1) Secant from 0.1% to 0.6% strain
(2) Calculated at the origin
(3) Secant from 0.01% to 0.1% strain
(4) Conditioned @ 77 ± 3 °C in 80%–85% R.H. for 20 ± 1 days

Table 7. Thornel T650-42/Commingled Radel 8320 composite properties

	Provisional Data Sheet				
Property	Unit	8-HS	4 × 4RH	Condition	Comment
0° tension (ASTM D-3039)					
strength	MPa	758	734	RTD	
modulus	GPa	76	74	RTD	(1)
0° compression (ASTM D-695 tabbed)					
strength	MPa	493	399	RTD	
modulus	GPa	–	72	RTD	(2)
0° compression (ASTM D-695 dogbone)					
strength	MPa	365	–	RTD	
modulus	GPa	81	–	RTD	(2)
± 45 deg tension (ASTM D-3518)					
strength	MPa	238	232	RTD	
	MPa	210	–	RTW	(4)
	MPa	–	195	105 °C wet	(4)
	MPa	128	–	163 °C wet	(4)
shear modulus	GPa	8	–	RTD	(3)
	GPa	6	–	RTW	(3) (4)
	GPa	–	6	105 °C wet	(3) (4)
	GPa	4	–	163 °C wet	(3) (4)
Open-hole compression (12 ply QI, 3.18 cm wide, 0.64 cm hole)					
strength	MPa	271	252	RTD	
	MPa	259	–	RTW	(4)
	MPa	240	–	105 °C dry	
	MPa	200	204	105 °C wet	(4)
	MPa	170	–	135 °C wet	(4)
	MPa	160	–	150 °C wet	(4)
	MPa	163	–	163 °C wet	(4)
Open-hole tension (12 ply QI, 3.18 cm wide, 0.64 cm hole)					
strength	MPa	369	354	RTD	
Quasi-Isotropic tension strength (12 ply QI, 3.18 cm dogbone)					
strength	MPa		664	RTD	
modulus	GPa		54	RTD	(1)
Quasi-Isotropic compression strength (12 ply QI, 3.18 cm dogbone)					
strength	MPa		363	RTD	
	MPa		295	105 °C wet	(4)
modulus	GPa		51	RTD	(2)
	GPa		51	105 °C wet	(2) (4)
Compression-after-impact (16 ply QI, 10.2 × 15.2 cm, 6673 J/m)					
strength	MPa	319		RTD	

(1) Secant from 0.1% to 0.6% strain
(2) Calculated at the origin
(3) Secant from 0.01% to 0.1% strain
(4) Conditioned @ 77 ± 3 °C in 80%–85% R.H. for 20 ± 1 days

10.6.2 Composite Laminate Properties

Composite laminate properties are presented in Tables 4 to 7 [45]. Property values reported reflect all the key tests as described in Chap. 7 and 9. The data base places a strong emphasis on compression properties both with and without holes or stress risers and in a post impact condition. The data clearly show the inherent damage tolerance achievable with thermoplastic matrix composites versus thermosetting based composites and excellent property retention under elevated temperature wet conditions.

Table 4 provides room temperature dry properties for standard modulus carbon composites based on fibers. Tables 5 to 7 consider properties for Radel 8320 composites reinforced with Thornel T650-42 intermediate modulus carbon fibers. The tensile properties decrease as one goes from Unitape (Table 5) to Interlaced tape (Table 6) to woven commingled fabric (Table 7). Properties are reflective of the composite construction i.e., the interlaced (cloth) based composites have approximately 50% of the fiber in the longitudinal direction versus unitape. Hence the tensile properties are nominally halved. Similar explanations apply to the components based on commingled fabric.

Off axis properties, each as $\pm 45°$ tension and open hole compression show very little dependency on prepreg product form as one might anticipate.

The compiled data discussed above are self consistent and are based on composites with a well defined processing or consolidation history. Lack of consistent processing history precludes an attempt to compare composites properties based on different matrix resins. However, Buchman and Isayev [46] present a good comparison of water sorption for composites based on several matrixes such as PEEK, PES, and PEI.

10.7 Summary

Thermoplastic matrix composites afford the potential of reduced manufacturing costs with the advantage of unlimited outlife and a wide variety of prepreg forms to enhance flexible processability. Composite laminate properties reflect inherent damage tolerance and excellent property retention under elevated temperature/wet conditions. An excellent data base is presented for Radel 8320 based composite system demonstrating outstanding produceability and a balanced property profile.

10.8 References

1. Anon, "The Place for Thermoplastic Composites in Structural Components," NMAB 434, National Academy Press, 1987.

2. S. Witzler, Advanced Composites, March/April, 1988, 55.
3. M. J. Michno, Jr., M. Matzner and G. T. Kwiatkowski, in "International Encyclopedia of Composites," Stuart Lee, editor, Vol. 4, p. 360 (1991).
4. S. Beland, "High Performance Thermoplastic Resins and Their Composites," Noyes Data Corp., Park Ridge, N.J. 1991.
5. B. Cole, Proc. 30th National SAMPE Symposium, 799 (1985).
6. J. G. Fitzpatrick, "Polyamide-imides (PAI)," 128, Engineered Materials Handbook, V. 2. Engr. Plastics, ASM (1988).
7. F. W. Harris, M. W. Beltz and P. M. Hergenrother, SAMPE J., 24, 1, 6 (1988).
8. N. J. Johnston and St. T. L. Clair, ibid., 24, 1, 12 (1988).
9. A. Yamaguchi and M. Ohta, ibid., 24, 1, 28 (1988).
10. Progar, D., et al., ibid., 26, 1, 28 (1990).
11. M. Molyneux, Composites, 14, 2, 87 (1983).
12. W. J. Lee, J. C. Seferis and D. C. Bonner, SAMPE Quarterly, 17, 2, 58 (1986).
13. A. J. MacGowan, W. C. Crowe and A. J. Barnes, Proceedings Tenth Thermoplastic Matrix and Low Cost Composites Review, USAF, Feb. 1991.
14. S. Clemans and T. Hartness, SAMPE Quarterly, 20, 4, 38 (1989).
15. J. Muzzy et al., J. SAMPE J 5, 15 (1989).
16. M. E. Ketterer, NASA Contractor Report 3849, 1984.
17. T. Lynch, SAMPE J., 25, 1, 17 (1989).
18. J. Vogelsang et al., CT1 39/91, 244 (1989).
19. A. P. Majidi et al., SAMPE J. 24, 1, 12 (1988).
20. Anon, Plastics Technology, November 1986, 13.
21. J. R. Koone, and J. H. Walker, ibid November 1986, 61.
22. R. S. Dave, A. Mallow, J. L. Kardos and M. P. Duderkovic, SAMPE J., 26, 3, 31 (1990).
23. F. N. Cogswell and D. C. Leach, SAMPE J., 24, 3, 11 (1988).
24. W. Sall and T. G. Gutowski, ibid., 24, 3, 15 (1988).
25. J. Muzzy, L. Narpoth and B. Varughese, ibid., 25, 1, 23 (1989).
26. S. Witzler, Advanced Composites, Jan/Feb., 53 (1987).
27. J. S. Colton et al., SAMPE J, 23, 4, 19 (1987).
28. S. M. Grove, Composites, 19, 5, 367 (1988).
29. B. J. Anderson and J. S. Colton, SAMPE J, 25, 5, 22 (1989).
30. R. B. Ostrom, S. B. Koch and D. L. Wirz-Sajranek, SAMPE Quarterly, 21, 1, 39 (1989).
31. S. Witzler, Advanced Composites, Sept/Oct., 49 (1988).
32. J. A. Sneller, Modern Plastics, 44 (1985).
33. R. J. Philpot, D. K. Buchmiller and R. T. Barker, SAMPE J., 25, 5, 9 (1989).
34. J. L. Kittleson, ibid., 26, 1, 37 (1990).
35. A. R. Madenjani, et al., SAMPE Quarterly, 16, 2, 27 (1985).
36. L. Leonard, Advanced Composites, July/Aug., 28 (1988).
37. A. Benatar and T. G. Gutowski, SAMPE Quarterly, 18, 1, 34 (1986).
38. E. M. Silverman and R. A. Griese, SAMPE J., 25, 5, 34 (1989).
39. R. C. Don et al., ibid., 26, 1, 59 (1990).
40. A. B. Strong, SAMPE Quarterly, 21, 2, 36 (1990).
41. P. Vautey, ibid., 21, 2, 23 (1990).
42. B. T. Astrom and R. B. Pipes, ibid., 22, 2, 55 (1991).
43. T. G. Gutowski, R. Henderson and C. Shipp, SAMPE J., 27, 3, 37 (1991).
44. A. Buchman and A. I. Isayev, ibid., 27, 4, 19 (1991).
45. Amoco Performance Products, Inc. Provisional Data Sheets.
46. A. Buchman and A. I. Isayev, SAMPE J., 27, 4, 30, (1991).

11 Applications

11.1 Aircraft

11.1.1 Structural

The aerospace or aircraft industry is international in scope [1]. Commercial and military applications utilizing ACM have been worldwide with the U.S. overshadowing the rest of the world. As an industry it consumes the largest quantity of ACM and is expected to continue through the end of the century. A Boeing spokesman has claimed that composites could be a large portion of the operating weight of commercial aircraft by 2000. Correspondingly smaller civilian aircraft progress has been quite rapid in the use of ACM. The Federal Aviation Administration (FAA, U.S.) has certified the first all composite business aircraft, the Beech Starship I (Fig. 1).

ACM provide weight reductions of 25 to 40% in the replacement of aluminum. Weight savings and the resulting dollar value may range from $50 to $500 per kg of weight saved but is also related to aircraft type/requirements and prevailing cost of fuel.

The use of ACM in aircraft has been methodical with applications noncritical to flight being considered initially. These nonstructural components included spoilers, elevators, horizontal stabilizer boxes. The success of these parts spawned efforts in structural segments of the aircraft such as tail sections (Airbus), rudder (Boeing 767), fuselage (V22), and wing (Prototype ATF, V22). In some of the large commercial aircrafts, ACM can be as high as 15% in the A320 Airbus (Fig. 2) or 6100 kg while the Boeing 767 (Fig. 3) contains 3% ACM. The large military C-17 cargo transport has about 6% ACM or approximately 7000 kg.

Military aircraft contain a higher ACM quantity since special military requirements consider payload, mission, and radar profile. The revolutionary V22 Osprey (Fig. 4) or tilt rotor helicopter is over 50% ACM with a total weight of about 15 000 kg. This represents a 20% weight reduction and corresponding 85% reduction in parts. The all composite helicopter, BK 117 (Fig. 5) by MBB, West Germany, and LHX (U.S. Army Light Helicopter) are further examples of how ACM technology is being driven by the needs of the military. For example the BK 117 has an all composite airframe and the complete ACM structure consists of 75% CF, 22% aramid and 3% glass. It also has a sevenfold reduction in part count from 745 (metal equivalent) to 105.

Significant improvements in resins, fibers and processing have facilitated the displacement of metals by ACM. The remarkable speed in commercializing high

Fig. 1. Beech Starship

Fig. 2. Airbus A320, courtesy of Airbus Industrie

Fig. 3. Boeing 767

Fig. 4. V-22

Fig. 5. MBB, BK 117

strength/high modulus carbon fiber has resulted in many opportunities for CF composites in critical structural areas such as fuselage, tail sections, and wing components or for the all composite aircraft. Newer aircraft designs with higher speeds have resulted in aircraft with higher skin temperatures, hence the need for higher T_g resins such as Bismaleimides and Polyimides. These newly developed thermoset resins upon modification are improved in impact. In the higher heat regions of the aircraft, PMR-15 is utilized as the matrix resin with carbon fiber for engine ducts, nacelles. Newer high T_g thermoplastics, particularly Radel, Torlon, PEEK and PPS, are undergoing extensive evaluation in many structural and non-structural aircraft component areas because of the greater toughness of these thermoplastic resins. Furthermore these thermoplastic resins may be the only matrix resins that will allow the new intermediate modulus, high strength carbon fiber, e.g., T650-42 and IM-8, to utilize its full strain potential in composites. Further short fiber and long fiber filled thermoplastic resins are being considered for several molded parts that range from control levers, instrument panels, radomes, even window frames.

To approach high utilization levels of ACM, progress must be made in several areas:

1. Substantial reduction in fabrication cost where, in fact, it is estimated to be currently over 70% of cost of producing finished components from composite. Thermoplastic resins may potentially reduce the fabrication cost.
2. Increased use of automation-significant cost reductions can result in aircraft component reduction via automated lay-up and parts inspection techniques.
3. Enhanced durability particularly for components in which damage may not be readily inspectable.
4. Ease of repair/repairability–especially important to develop techniques and materials which minimize downtime.
5. Decreased risk via increased data base and flight experience.

Coupled to these problem areas are newer challenges that increase property requirements of ACM for newer aircraft under consideration such as Mach 3 passenger aircraft or improved Concord with Mach 2-3 capability and the National Aerospace Plane (NASP) with a proposed Mach 2.5 capability.

11.1.2 Aircraft Interior Components

Federally mandated regulations issued by the U.S. Federal Aviation Administration (FAA) provide for enhanced fireworthiness of commercial aircraft interiors. Cargo liners as well as large interior panels are required to pass stringent flammability tests. The purpose of the regulations is to allow time for passenger egress prior to flashover or when explosive propagation of fire makes cabin interiors nonsurvivable. For aircraft interior panels, the Ohio State University (OSU) heat release test (see Analysis/Testing section) measures peak heat and total heat release during a 5 minute sample burn. Typically aircraft interior panels are non-load bearing structures whose surfaces are covered with decorative laminates or protective coatings. These interiors consist of sidewalls, stowage bin doors, cabin partitions, galley and lavatory walls and ceiling panels (Fig. 6).

a) Interior Panels

Early enactment of the regulation in August 1988 required that all newly certified aircraft conform to new flammability standards of 100 kW/m^2 peak heat release and 100 kW-min/m^2 total heat release. In August 1990 all aircraft in various stages of manufacture required upgraded interiors. Also, large aircraft certified after 1958 must retrofit with interiors meeting the new regulation of peak heat release of 65 kW/m^2 and total heat release of 65 kW-min/m^2.

Studies conducted by Nollen [2] focussed on the total composition of large interior aircraft panels such as decorative laminate, resin, fiber reinforcing

Fig. 6. Aircraft interior, courtesy of Airbus Industrie

materials, and honeycomb core. Heat release data for currently used epoxy resins versus phenolics indicated that little difference in heat release exists between the two resins especially at low resin content. As the resin content increases, the combustibility of epoxy resins becomes apparent with the epoxy panel emitting more heat than the phenolic panel. These differences become more evident as one compares the low smoke evolution of phenolics versus dense smoke emissions of epoxies. To use phenolics as replacements for epoxy resins for aircraft interiors, tack/drape, and vacuum bag characteristics are necessary to integrate phenolics into existing prepreg operations. In many cases modified phenolics (thermoplastic resin modified) are used in epoxy replacement systems. Many companies have developed modified phenolic resins and include SP Systems, Fiberite, American Cyanamid, Hexcel and Heath-Techna [3]. For non-contoured interior parts, crushed-core systems of unmodified phenolic to honeycomb are used to enhance peel strength and flow. With increased pressure during cure a more intimate bond is formed between the phenolic reinforced fabric prepreg and honeycomb core. A recent phenolic resin known as Temephen D50 [4] has been reported to be effective in crushed core panel.

b) Cargo Liners

Cargo liners are usually monolithic laminates of one to a few plies of cured prepregs (Fig. 7). Materials consist of aluminum, E or S-2 Glass, or Kevlar with epoxy, polyester and more recently phenolic resin. Some cargo liners contain honeycomb sandwich panels. The resulting liners are fastened onto metal frames. Thick liners are used on side walls or areas of abuse while thinner liners are satisfactory for ceilings. Most liners are covered with a decorative material such as polyvinyl fluoride (Tedlar, Dupont) film for ease of cleaning and maintenance.

Fig. 7. Cargo liners

In 1986 the FAA issued a new rule which upgraded burn through resistance of Class C and D cargo liners for commercial transport aircraft. Class C and D cargo liners are used in large compartments (6–45 m^3) that are inaccessible during flight. These compartments may contain fire detection and suppression equipment. The majority of cargo compartments in commercial transport aircraft are either Class C or D. The regulation is based on cargo liner ceilings and sideways surviving a 5 minute oil burner test with no burn through and back side temperature of less than 204 °C. The test consists of a two gallon/hour oil burner impinging directly on ceiling panel and indirectly on a side wall panel. The burner is maintained at 927 °C with a 9 watts/cm^2 heat flux. Thus all new aircraft certified after June 1986 have conformed to the new regulation. The FAA may also extend the rule to existing aircraft certified prior to June 1986.

Besides flame containment, other cargo liner attributes include light weight, abrasion resistance, resistance to fastener pullout, ease of cleaning and damage tolerance. The latter is not only important in service life but also for safety. Punctured or torn cargo liners are not very efficient flame barriers. Cargo liners must also pass special impact tests that are specific to aircraft manufacturers (Boeing, McDonnell Douglas). Cargo liners based on modified phenolic resin systems combined with either S-2 Glass or Kevlar are supplied by M. G. Gill, Fiberite and Permali.

11.1.3 Aircraft Brakes

As civilian and military aircraft became heavier and faster, improved aircraft brakes became necessary. The high heat developed during normal or aborted stops has required the use of carbon–carbon composites which possess necessary high heat and thermal shock resistance, excellent friction and wear properties. Carbon-carbon brakes (Fig. 8) that possess these characteristics provide superior stopping capability, survive many abortive stops, and endure

Fig. 8. Aircraft brakes

much longer than earlier aircraft brakes composed of steel or copper. The disk braking mechanism generates considerable frictional heat which must be dissipated. Besides heat resistance and high temperature stability, C/C composites have low thermal expansion, good thermal conductivity, and a favorable friction coefficient independent of temperature. C/C's ability to slide against itself without galling, makes the brakes highly wear resistant, resulting in longer times between repairs. The CTE of C/C composites remains somewhat constant over a broad temperature range. C/C has a high heat capacity, so it can act as a lightweight heat sink. C/C is less susceptible to warpage than metal. The low density value of 1900 kg/m^3 for C/C composites represents nearly four times the braking power of the forerunner steel brakes.

Manufacturers of C/C composite aircraft brakes combine CF, consisting of PAN T-300 or pitch P-25, with phenolic resin, or liquid pitch impregnation, carbonize, and chemical vapor densification treatment to the C/C composite. The multi-step, labor and energy intensive process contributes to the relatively high acquisition costs. However the life cycle of the C/C composite systems prove very effective based on weight savings and life expectancy of two to four times that of steel/cermet brakes. C/C composites are the optimum braking material for civilian and military aircraft. The largest volume usage of C/C composites is for the manufacture of aircraft brakes.

11.2 Ballistics

ACM plays an important role in new and improved ballistic components. The replacement of metal particularly steel and aluminum has resulted in improved ballistic protection with a corresponding reduction in weight. Key application areas consist of personal protection, land vehicles and shipboard use. High performance composites for ballisitic applications are prepared by combining Kevlar 29, S-2 Glass, or Spectra (UHMWPE) fibers with a high modulus thermoset resin system. Some low modulus thermoplastic resins have been combined with Spectra fiber. The resulting ballistic components are quite flexible and are utilized with other rigid ballistic composites (ceramics, spall liners).

11.2.1 Components for Ballistic Components

Typical ballistic components are listed in Table 1. These composites are relatively low resin matrix containing composites which rely on the strength/modulus characteristics of the fiber for ballistic performance. For ballistic evaluation or comparison, a value known as ballistic limit (V_{50}) and expressed as velocity is determined. The ballistic limit is the approximate velocity at which 50% of the projectile impacts penetrate completely and 50% are partial penetrations (Military Standard 662D, U.S. Army). The V_{50} determination allows a comparison of the efficiency of ballistic materials and is related to composite areal density or weight/unit area (kg/m^2). Various size projectiles are evaluated and range from 0.22, 0.30 to 0.50 caliber (5.59, 7.62, 12.7 mm).

In 1978, the US Army adopted the Kevlar composite helmet, Personnel Armor System-Ground Troops (PASGT) (Fig. 9). It replaced the Hadman or M1 helmet which consisted of a steel outer shell with a nylon resin treated liner. In excess of 2 million Kevlar helmets have been manufactured to date. Devils Lake Sioux Manufacturing Corporation of Fort Totten, N.D. is the largest manufacturer of the Army helmet and is responsible for producing over a million helmets (Table 2).

Similarly flat plate armor based on Kevlar 29, "E" or S-2 Glass composites for personnel vehicles or tanks continues to replace steel or aluminum. A ballistic liner consisting of a Kevlar composite is being installed in critical areas inside the U.S. Army M113 armored personnel carrier. These liners are also planned for the M2/M3 Bradley fighting vehicles. S-2 Glass fiber composite is also being examined for the M113 as well as the Bradley Turret and the Army's HUMMER (High Mobility Multipurpose Wheeled Vehicle).

These composites of Kevlar 29, E or S-2 Glass can function as spall liners for ceramic armor materials which provide better protection for combat vehicles

Table 1. Components for ballistic composites

Fibers:	Kevlar 29, S-2 Glass, Spectra 1000
Matrix resins:	Vinyl esters, polyesters, phenolics and modified phenolics
Weight per cent resin to fiber:	15–20: 85–80

Fig. 9. Ballistic helmet

Table 2. Comparison of U.S. Army helmet materials[a]

Materials	V_{50} (Ballistic limit)[b]
Steel/Nylon liner	442 m/sec
Kevlar composite	610 m/sec

[a] 1.59 kg; 10.74 kg/m^2 areal density
[b] 0.22 caliber, 17 grain fragment simulating projectile

against medium and heavy threats. Figure 10 shows the utility of these spall liners with ceramic materials.

With Kevlar or Spectra, ballistic performance is mainly related to fiber strength. Ballistic impact energy is dissipated in a direction perpendicular to the projectile path [5, 6]. Some energy is also dissipated in the transverse direction parallel to the projectile path. Either fiber is able to spread the impact energy throughout the fiber network. An entirely different mechanism of ballistic impact occurs with S-2 Glass composites. Bless and Hartman [7] have reported that a positive relationship exists between projectile diameter and composite areal density. Further they have shown that V_{50} is proportional to composite transverse compressive strength. Thus high strength organic fibers (Kevlar, Spectra) provide ballistic performance due to fiber strength whereas the ballistic performance of S-2 Glass is related to composite properties.

A new improved ballistic method has been recently disclosed [8] and is reported to increase ballistic values (V_{50}) from 10–25% with 0.22 and 0.30 caliber projectiles for areal densities of 12.2–19.5 kg/m^2 as compared to similar areal densities of Kevlar composite controls.

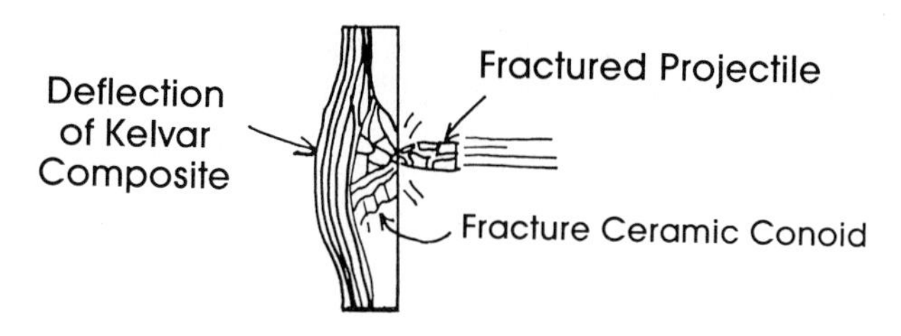

Fig. 10. Ceramic armor

Several opportunities for ACM in ballistic applications include:

1. Retrofitting and armor improvement for helicopters and fixed wing aircraft;
2. improved armor for various land vehicles; and
3. improved ballistic protection aboard naval vessels such as critical above-deck areas.

Improved structure, design and ballistic performance are achieved in the fabrication of the Bradley Turret with ACM. It is based on an S-2 Glass polyester composition that results in a 15% weight reduction for the composite turret with similar ballistics as its metal replacement. Rapid rotating speed and a wider and more effective line of fire are key features of the lighter composite turret. Following-up the successful development of the composite turret, the Composite Infantry Fighting Vehicle (CIFV) used a Bradley Fighting Vehicle chassis (Fig. 11). The composite hull is mainly a floor deck and two side panels, each 2.4 m $\times$ 6.7 m with a thickness of 22 to 69 plies of S-2/polyester resin. Each panel weighs 820 kg after machining. Fabrication improvements have allowed a

Fig. 11. CIFV

25% weight savings for the armored hull of the CIFV. Additional benefits of the composite armor structure consist of better insulation, lower radar and thermal signatures, reduced vibration, less noise. Modification of the US Army M939 5 ton truck via a composite armor approach is being considered.

Similarly there are opportunities for composite armor aboard naval vessels especially for the protection of ammunition and missiles that are stored on deck from air attack. Even the protection of ammunition in the ship's hold is a consideration.

11.2.2 Future Opportunities for Ballistic Composites

The integration of some of the above key armor composite developments is fostering the consolidation of several critical functions within the same composite. These functions are structural strength, ballistic strength and invisibility to radar. Newer armor composites will attempt to incorporate these functions into the armor. Fiber strength (Kevlar, Spectra) and composite strength (S-2 Glass) for improved ballistics coupled to the strength/modulus characteristics of carbon fiber as well as woven hybrid fiber systems will be important material considerations for future armor composites.

11.3 Space

The use of ACM for space applications involves two distinct areas: launch systems and self-contained space modules.

11.3.1 Launch Systems

The launching of any object into space must consider the weight of the lift vehicle, payload in orbit, expendable or recoverable launch system, cost of payload per kg, liquid or solid propellant, ACM or metals. Current launch vehicle technology and hardware span several decades. Efforts to revamp or upgrade vehicles, support vehicles, support operations and various manufacturing processes to achieve lower costs, higher payloads, reliability and more frequent launches have prompted NASA/U.S. Air Force to initiate an advanced launch system (ALS) program which will provide improved launching technology by the turn of the century.

ACM are preferred over metals (allowing a 30% weight savings) for expendable launch vehicles (ELV) for Atlas, Delta, and Titan rocket launch systems. Use of composites for expendable systems can be cost effective since ELVs require fewer parts due to automated manufacturing techniques like filament winding and automated tape winding which are more economical than the labor intensive joint welding of metals. Thiokol (U.S.) has a long history of involvement with the fabrication of metal solid rocket motor cases and is a leader in composite cases. Composite motor cases have evolved from S-2 glass/epoxy to aramid/epoxy to the current standard CF/epoxy with the latter providing the cases with necessary structural stiffness and strength to carry loads with minimum weight.

Thiokol has established a leadership position in using prepreg tow (Amoco T-40/ERL 1908) to wind solid fuel cases (e.g., SICBM) which exhibit exceptionally high uniform performance (low coefficients of variation, CV's $< 1\%$). Filament winding (primarily wet winding technology) of CF/epoxy has been developed by Hercules Aerospace (U.S.) for solid rocket motor cases as boosters for Delta and Titan launch vehicles which rely mainly on liquid propellant core engines. Use of these boosters has increased payload depending on whether low earth orbit (LEO) or geosynchronous earth orbit (GEO) is considered.

Thermoplastic resins (PEEK) are being examined with CF in the fabrication of filament wound pressure vessels for evaluation as liquid propellant tanks (O_2 or H_2). Preliminary testing indicates thermoplastic filament wound cases resist microcracking and maintain mechanical properties after severe and prolonged testing.

Nozzles such as exit cone and throat elements are comprised of ACM of phenolic/CF or carbon–carbon composites to withstand hot exhaust gases of burning propellant (up to 2700 °C) and maintain shape.

11.3.2 Self-Contained Space Modules

A variety of space modules are available and range from satellites for communications, space craft for space exploration, space stations (platforms and Space Shuttle Orbiter). The key factors for these exotic, futuristic probes are weight and survivability.

11.3.2.1 Commercial and Military Satellites

Spin stabilized satellites are being displaced by a new generation of body stabilized systems. New materials requirements according to Hughes aircraft (U.S.) consist of higher strength and stiffness components with greater thermal/electrical conductivity resulting in greater use of composites. Use of high modulus composites based on pitch fibers [9] (Amoco P-75, P-100 and P-120) has been advanced by Amoco's pioneering development of thin (1 mm/ply cured thickness) prepreg which affords optimization of component stiffness and strength at minimum gauge and weight. In addition the newest high performance pitch fibers (Thornel Hi-K, Amoco) are being evaluated in satellites for thermal management in satellite radiators and solar arrays. These pitch fiber based composite systems have three to four times the thermal conductivity of copper.

11.3.2.2 Spacecraft

Space probes such as Magellan and Galileo are low mass objects allowing more scientific instrumentation to be incorporated into the payload. The Magellan is streaking towards Venus while the Galileo was launched recently towards Jupiter. Both of these interplanetary travelers will gather further information of their target planet prior to their departure into deep space.

An unusually large spacecraft that is maintained in orbit 500 km above the earth is the Hubble Space Telescope (HST) (Fig. 12). The HST weighs 11.6 tons (comparable to a railroad boxcar) and is the largest probe launched into space. For precise alignment of the exotic optics aboard the Hubble, a metering truss of CF/epoxy provided the dimensional stability required with near zero coefficient of thermal expansion (CTE). The critically low CTE virtually excluded metals from support systems module equipment. Trusses for all these spacecraft are thin wall CF/epoxy tubes calculated to carry various compression loads. Besides HST, other NASA Great Observatory series of astronomy satellites are the Gamma Ray Observatory (GRO), the Advanced X-ray Astrophysics Facility (AXAF), and the Space Infrared Telescope Facility (SIRTF). It is hoped that all four observatories will operate simultaneously and allow some important sporadic events such as novas or possibly a supernova to be observed with unprecedented clarity in all four wavebands of the respective observatories.

11.3.2.3 Space Stations

The proposed U.S. Space Station, Freedom, will rely heavily on ACM. The trusswork of the two primary station booms provides a lattice network ranging from the solar arrays at each end to the central crew cabins and lab modules.

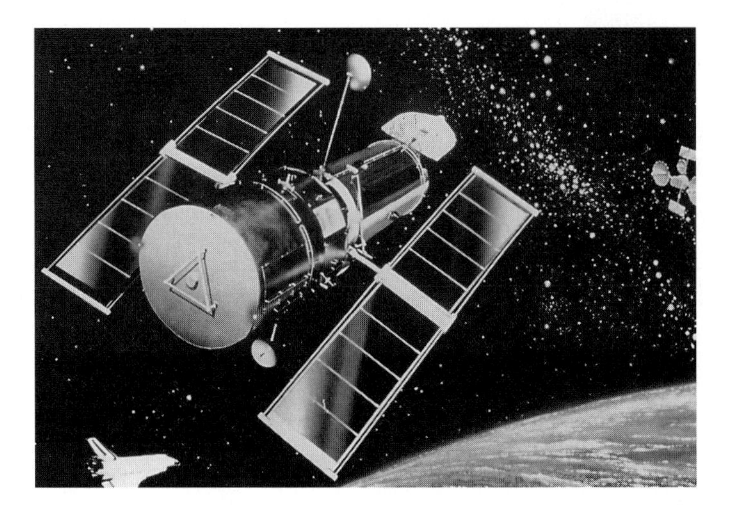

Fig. 12. Hubble space telescope

The truss is composed of 500 to 1000 tubes made from 40 million modulus CF with toughened epoxy that resists microcracking at cryogenic temperatures. The favorable features of ACM (weight, cost, low CTE) in relation to aluminum or titanium were decidedly in favor of ACM. Prototype pressure vessels to propel the station into different orbits at various intervals are being developed. Tanks are CF/epoxy overwrapped on a seamless aluminum alloy liner. Burst testing of the tanks has shown that the tanks are 1.7 times more efficient than the all titanium spherical tank. Further testing is planned with oxygen and hydrogen at 6.89 MPa with tanks 200 cm long and 51 cm in diameter.

11.3.2.4 Space Shuttle

The first and significantly large ACM parts on the Space Shuttle (Fig. 13) are the payload bay doors made of CF/epoxy (T-300). The composite part represented a weight savings of over 400 kg and was directly transferable to bigger payloads. More changes of metals to ACM are contemplated with possibly retrofitting current orbital vehicles with CF systems with either bismaleimides or polyimides.

Fig. 13. Space shuttle

11.3.3 Future Opportunities in Space

Future applications of ACM to space hardware will be based on improvements in materials as well as improved manufacturing technology. An example of the latter is precision mandrel-wound, thermoplastic-matrix composites resulting in close tolerance control of tubular structural elements. Improved materials for application to space hardware include the use of composites based on Amoco's Thornel Hi-K pitch fiber to achieve higher thermal conductivity than metals at a fraction of the weight. Thin prepreg (Amoco's 1 mm/ply cured thickness) will be important in achieving optimal structural properties including dimensional stability and microcrack resistance under the thermal cycles inherent to orbital travel in space.

A current, focused study is improved structural properties of C–C composites to achieve parts consolidation, weight reduction, enhanced toughness and oxidation resistance. The use of coatings is an important aspect of improved C–C technology and has high potential for replacement of aluminum honeycomb structure with enhanced resistance to atomic oxygen and other radiation sources.

11.4 Sports/Leisure

Progress continues unabated in the application of ACM to sports equipment (Fig. 14). Beginners or professional athletes alike are demanding the best sports equipment that will perform better than previous models and also exhibit better durability. Psychologically the athlete expects to perform better in the sport

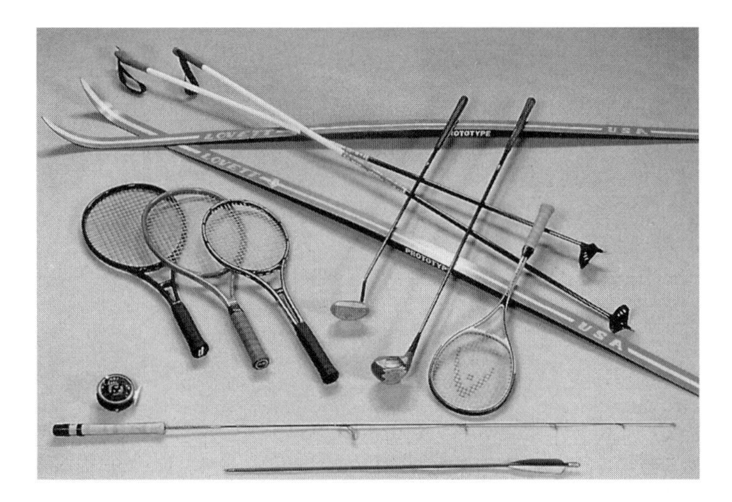

Fig. 14. Sports equipment

because of the newest and latest model. Sports equipment manufacturers have relied on ACM to achieve design flexibility resulting in sports hardware which is safer, stronger, stiffer, lighter, more durable and better able to dampen vibrations. The major advantage of ACM over metal or wood is that with ACM the specific properties can be designed to meet specific requirements. Equipment can be tailored to meet optimum performance criteria. Strength and stiffness can be introduced where necessary and in the desired direction. CAD and aerospace technology have helped to create a vast array of high technology sports equipment that clearly rivals all the early equipment models that were based on wood or metal. Form and function of the sports participant and his equipment are carefully scrutinized by the designer and the material scientist resulting in continuing innovation in sports equipment. ACM has provided more power to women in sports equipment particularly when CF is used in golf clubs, archery bows, tennis rackets, fishing rods.

Some sports items are quite labor intensive with many off shore manufacturing locations in the Pacific Rim area or the Caribbean. Some sports equipment manufacturers are multiple sports item manufacturers.

11.4.1 Golf

The use of ACM in the manufacture of golf club shafts and club heads has been aided by analyzing the design and material requirements of the club, weight distribution in the club, and the velocity at impact of the golf ball. Two key elements that have been evaluated in the design of new clubs are "swing and strike". A lighter shaft, tubular in structure with strong, stiff characteristics allows a rapid swing. A light club shaft permits more weight in the club head and thus obtain a longer driving distance due to enhanced momentum transfer from heavier club head to ball. ACM golf club shafts offer added distance, less fatigue and greater accuracy over metal shafts. Newer features of golf club shafts are intimately related to processing and materials. Most CF/epoxy shafts are made with medium and high modulus CF to control the complex dynamics (e.g., torsional stiffness) that result during swing and strike. New CF/epoxy shafts by Uni Fiber Corp. (San Diego, U.S.) are manufactured with high strength/high modulus CF (Amoco's T650-42) which provides excellent interface within the CF/epoxy laminate and distributes a significant increase in strength at the tip of the shaft. Novel processing by Uni Fiber permits different fiber orientation in each shaft comprised of parallel sections from the top to the bottom with a tapered middle section. The shaft's tapered section is affected by flex point and torque. The flex point controls trajectory of the golf ball: lower flex point results in high trajectory while high flex point yields a lower trajectory. Flex point is precisely matched at 40% of overall shaft length from the tip of the club. While club lengths and precise location of flex point on each club may be different, the flex point percentage is maintained throughout the set resulting in the same feel for all the clubs.

11.4.2 Tennis Rackets

This sport like golf involves ball speed and the participant's control of the tennis racket. It is a contact sport – the tennis racket making contact with the tennis ball. A rapid evolution of racket materials has occurred with the high performance composite racket displacing wood and metal rackets. The key features of the composite racket are high strength and stiffness in compression and bending mode of operation during game conditions. A combination of braiding and unidirectional CF fabrication results in a racket which resists twisting and has high stiffness. The matrix resin must sustain the high localized strains induced under impact conditions. Newer modified epoxies and thermoplastic resins afford the opportunity to further enhance performance under high impact loads.

Equipment manufacturers tailor the properties of rackets by designing a variety of features into the racket. Fibers such as CF, Boron, aramid, and Spectra, and hybrids thereof are used. The strength/stiffness of CF provides controllable power; structural stiffness reduces racket head deflection so that transmitted energy allows the ball to return with true angle. Aramid or Spectra permit a weight reduction and dampen vibration to prevent "tennis elbow".

11.4.3 Bicycles

Wind resistance is a major obstacle that must be overcome by the bicyclist when he is leisurely or laboriously peddling his bicycle (Fig. 15). Like aircraft, bicycles are tested in wind tunnels. It is estimated that up to 90% of the total force opposing a cyclist is wind resistance. Besides wind resistance, turbulence created by spokes, similar to an egg beater, has been identified. Spokeless or disc wheels

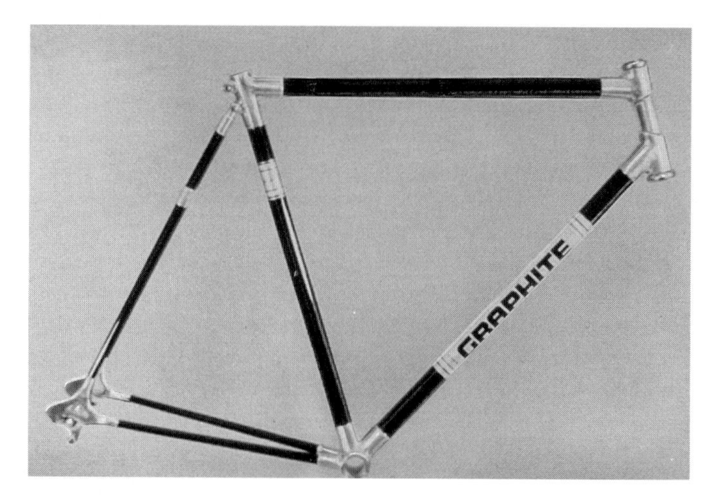

Fig. 15. Bicycle frame

and 3-spoke wheels based on ACM's have been introduced and lessen air drag. Although a majority of bicycles have metal (steel, Al, Ti) frames, new innovations of frames and materials are constantly occurring. Newer ACM frames can vary from a monocque design that is lighter and more aerodynamic in performance, to the use of CF braiding for high strength and resistance to torsion/twisting, or a one piece ACM tubular frame composed of materials ranging from CF (Amoco) to CF/Spectra hybrid. The CF is responsible for stiffness while the CF/Spectra hybrid allows a smoother ride attributable to the presence of Spectra for shock dampening as well as weight reduction.

An all purpose bicycle weighs about 14 kg whereas racing bikes weigh about 10 kg. For each kg reduced, price of the racing bike increases $200 in price. In excess of 10 million bicycles are sold in the U.S. To stimulate sales and interest, manufacturers continue to rely on new high technology products and components (frames, wheels brakes, seat, peddles) with the price of some light weight racing bikes exceeding $3000.

11.4.4 Skis

Skiing, particularly downhill, is a high speed sport especially demanding on equipment. For high velocity skiing, with speeds up to 200 km/h, high strength, light weight materials are highly desireable. The high stiffness of CF/epoxy skis and poles permits sharper, faster turns as well as speed. ACMs allow ski designers to have torsional and flexural stiffness while enhancing toughness or durability and thus overcoming the limitation of wooden ski construction. Some ACM skis have a layer of aramid which provides abrasion resistance, dampens vibration and hinders torsion. S-glass/thermoset resin system can be braided via an automated process into skis that are uniform without seams or overlap. Furthermore they are lighter than prepreg assembled skis due to closer tolerance and controlled resin content.

Conventional skis begin with separate pieces for the top, sides and bottom that are all joined together to form a box-like ski structure. A recently innovative ski design that relies on ACM improves stability and turning. The novel technique consists of fabricating the ski's top and sides into a single structure (based on Glass/epoxy) called a "one-piece cap" and likened to a bridge or an arch. As a bridge or an arch, pressure applied to any portion of the structure is rapidly transferred to the extreme edges of the ski. These innovative new skis hold their edges on snow better, turn quickly, easily and are more stable. Within the one piece cap is an inner core that is the main structural foundation of the ski providing flex distribution, torsional rigidity and weight.

The ski pole has also been improved in composition and performance. A key factor that is important to the skier is the swing weight of the ski pole. In composite poles, the center of gravity can be moved closer to the hand grip with a corresponding reduction in the swing weight. Besides swing weight, stiffness, compressive strength and durability are important. CF/epoxy (Amoco) ski poles exhibit significantly improved stiffness and compressive strength over aluminium or wooden poles. Unfortunately composite poles suffer in impact and lack in durability. Newer modified epoxies or thermoplastic resins are expected to improve impact when either resin system is used as matrix resin with CF.

11.4.5 Fishing Rods

The requirements of an optimum fishing rod are identified with improved stiffness/strength, lighter weight, better control of the lure, and ability to bend to almost the breaking point (360°) with no appreciable fatigue. It must endure stress while bending especially if a heavy fish is engaged. Hybrids of CF/glass with epoxy result in ultralight rods that allow one to cast farther, make more casts without becoming arm weary, and yet afford the strength to haul in the big ones. Use of ACM has provided the designer with the ability to tailor rod torsional and flexural response resulting in both enhanced overall rod strength as well as enhanced rod tip sensitivity. High and intermediate modulus CF, such as Amoco Thornel T 650-42, have contributed significantly to the new high tech fishing rods.

11.4.6 Archery

This highly specialized sports activity is undergoing a materials transformation from wood or metal to ACM. High performance archery bows can be pultruded S-glass fiber or fabricated from CF/thermoset prepreg. These high technology bows allow the archer to shoot arrows farther, faster with a high degree of accuracy. Accuracy is also attributed to new composite arrows of pultruded CF. The CF arrow has a smaller cross section which reduces surface area, reduces aerodynamic drag and minimizes crosswind effect. All these factors contribute to accuracy. Besides accuracy, the lighter, slimmer and faster ACM arrow shafts resist bending and shattering that usually occur with aluminum arrows.

11.4.7 Other Miscellaneous Sports

Smaller quantities of ACM are currently being utilized in canoes, kayaks (paddles), sleds, composite kites, sailboards, water skis, baseball bats, oars, hockey and lacrosse sticks, vaulting poles, table tennis bats, exercise rods.

11.4.8 Future Efforts in Sports Equipment

Modified epoxies and thermoplastic resins with selective fibers or hybrids will be examined for improved impact wherever required for the sports item as well as reduce or minimize microcracking of unmodified epoxies or polyester resin systems.

The role of thermoplastic resins as matrix resins may be one of the factors for some sports equipment manufacturers to return to the U.S. Contemplated short fabrication cycles of thermoplastic resins versus lengthy cure of thermoset systems would be a contributing factor.

11.5 Naval Vessels

Early fiber reinforced pleasure marine vessels based on glass/polyester were manufactured in the 1940s and provided many benefits such as lighter weight, higher corrosion resistance and more economy of manufacture than the corresponding wood or metal vessels. The emergence of ACM as high performance materials has extended their use into high speed craft and deep ocean submersibles. The U.S. Navy has traditionally used steel, aluminum and glass for their vessels except for shipboard armor which is based on S-Glass and aramid composites.

Research programs are underway to assess the value of ACM in next generation surface ships and submarines. A complex set of requirements include stringent demands on flammability resistance and toxicology of combustion products as well as the ability to sustain the high compressive loading encountered in deep submergence. The compression loading for submarines is likely to require the ability to fabricate a thick wall structure which is resistant to local high impact and/or shock loading.

An equally demanding requirement characteristic of surface vessels is light weight structure with excellent damage tolerance with thin gauge, e.g., a deck house sidewall. The above water line structure weight savings affords enhanced stability and performance.

The fire/toxicity ACM requirements for submarine interiors are more severe than the newly enacted FAA requirements for aircraft interiors (see Aircraft Interior section); flammability standards are being developed by NIST and the U.S. Navy. The submarine pressure hull which is subjected to external hydrostatic compressive loads would require a low weight to displacement ratio for maximum payload while maintaining optimum range and speed. Clearly the choice of materials for the submarine hull to possess light weight, corrosion resistance, and high compressive strength is CF/epoxy composite. This long term effort is most challenging because the development of many new concepts in hull design, fabrication, and joining techniques is required to meet the goal of the program – reduce weight and improve submarine survivability.

Surface ships are undergoing a number of development studies with ACM and deal with weight reduction in the deckhouse, armor, and other areas. The aim is to develop a combatant deckhouse with improved survivability and reduce weight/maintenance costs through the use of ACM. The USS Essex,

Fig. 16. USS essex

damaged during Middle East duty in 1989, is the first Navy ship to have a composite armor deckhouse (Fig. 16). The armor is made of high performance S-2 glass fiber composite. The decision to use the fire resistant armor was based on favourable ballistic data and lower cost than conventional armor. Ease of fabrication and installation of S-2 composite were additional features when compared with conventional armor.

11.6 Tooling

In the fabrication of ACM parts, tools (Fig. 17) are necessary to shape or conform the intermediate composite into the desired configuration. Early tools for composite parts were heavy, imprecise and not very durable. The rapid growth of ACM, particularly for aerospace applications, created a need for

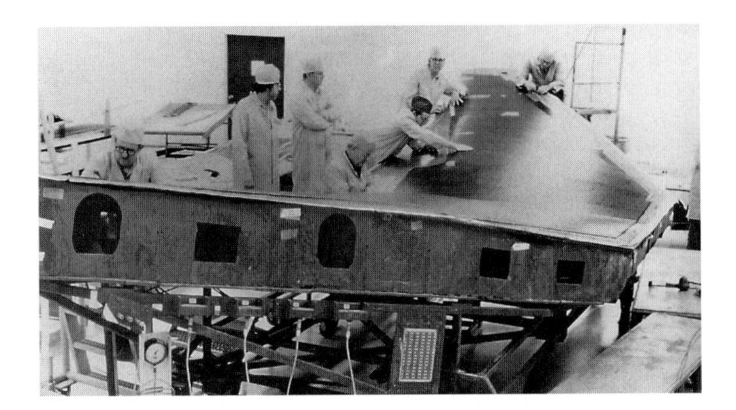

Fig. 17. Tool

Table 3. Properties of various tooling materials

Property	Monolithic Aluminum	Steel	Nickel	Graphite	CF/Epoxy Composite	Ceramic
Density g/cm^3	2.72	7.9	8.9	2.7	1.6	4.4
CTE (10^{-6}/°C)	24.5	11.0	13.3	3.1	1.8	3.6–7.2
Thermal cond. (W/mK)	202.4	45	58.8	117.6	< 8.5	28.4
Specific heat (J/kgK)	0.92	0.46	0.46	0.71	1.00	0.67

tooling materials more suitable than conventional steel and aluminum materials. Although steel tools are durable, surviving thousands of cycles, its bulk density makes it difficult to move, energy inefficient (heating/cooling cycles) and more importantly CTE differs (Table 3) radically from that of ACM, causing distortion of contoured parts during cure. Other comparisons of various tooling materials are listed in Table 3.

Metal tool materials are undesirable in spite of very favorable durability. Monolithic graphite with an attractive CTE appears to be a good tooling match for fabricating CF/epoxy parts. It requires costly machining, suffers from fragility and size limitations. Ceramic tools based on chemically bonded ceramics are new materials that offer great potential as ACM tools especially for high temperature thermoset or thermoplastic CF components. Cost, weight, and thermal mass of ceramic tools may limit full penetration into tooling of ACM parts.

The most important advantage of CF/epoxy composite tooling for making composite parts is the thermal similarity (matched CTE) that is crucial for unusually shaped, contoured parts. Hence low CTE tooling materials are desirable and consist of monolithic graphite, CF/epoxy composite and ceramic. The value of the tool is identified with tool maximum temperature and number of cycles that the tool can be successfully used. CF/epoxy composite tools perform best in a temperature range below 121 °C, marginally up to 177 °C while above 177° is undesirable. Industrial experience of CF/epoxy tools results in about 30–50 cycles at 177 °C. Above 177 °C, CF/epoxy has an abbreviated lifespan. Tool lifetime and durability are subject to many factors such as damage or physical abuse, improper cutting practices, microcracking or communication lapses between tool designers and fabricators. New tools based on CF/BMI with a tool temperature to 246 °C for PMR-15 parts and Dupont's Avimid N tool (CF/NR150 B2) have been introduced. The high cost of Avimid N Tool may limit its broad use in high temperature matrix resin systems.

11.6.1 Future Efforts in Tooling

Better tooling is still required for large, complex structures fabricated out of high temperature thermoset or thermoplastic resin systems. In addition to developing new matrix resins, current efforts focus on the design of woven fabrics with high yield carbon fiber tow to gain cost efficiencies in tool construction as well as enhanced performance.

11.7 Automotive

The automotive industry is quite cost sensitive in material utilization and, in most instances, these materials are fabricated into automotive components via automation with accompanying high production rates. Typical economic characteristics such as low cost, highly automated process, and high production rates are contrary to the general norm of ACM. ACM are generally more expensive, are fabricated in low production rates, and hence fabrication process have not required high automation. Nevertheless there are some limited automotive applications that are yielding to ACM.

Composite drive shafts (Fig. 18) based on CF/epoxy (CF stiffened aluminum shaft) for General Motors in selected light trucks and vans are approaching 500 000 medium size shafts (1.8 m × 10 cm diameter) and represent a respectable volume of CF, nearly 230 tons/year.

An early developmental application involves a mostly composite engine of a moldable glass phenolic composition. The composite internal combustion engine known as "Polimotor 234" is a 2.3 liter, 4 cylinder engine block that is molded and ready for use in about 4 minutes. The elimination of extensive machining reduces the cost by 75% over the customary metal engine production. Weighing about 80 kg it delivers 175 HP as compared to 160 HP for the cast-iron counterpart that is 50 kg heavier or 130 kg total weight. The ACM reduces engine weight, size and noise while improving gas mileage. During engine operation the composite engine is exposed to temperatures varying between 125 to 205 °C as well as bolt loads that can exceed 69 MPa.

A composite PPS-steel camshaft is being developed for selected General Motors production vehicles. The PPS-steel composite is claimed to offer several advantages over steel shafts including a significant cost reduction, 50% weight saving, less engine noise, and a manufacturing cycle shortened from hours to just 40 seconds. Glass reinforced PPS is injected into the voids around the cam elements that are positioned on a splined shaft. The reinforced PPS provides sufficient high temperature resistance (260 °C), dimensional stability and chemical resistance. During operation the PPS is subjected to 7.6 MPa compressive stress and an operating temperature range of 125–150 °C generated by engine oil and oil additives.

Fig. 18. Composite drive shaft

Connecting rods (conrods) and piston rings of CFRP with a variety of resins within a temperature range of 130 to 180 °C have been described by Beckmann and Oetting [10]. No damage was detected when redesigned CFRP conrods were examined under full load to a fatigue limit of 20×10^6 cycles. Tests are in progress with these designs being adapted to a geometrically compatible engine. CFRP rotating piston rings were unsuccessful metal replacements due to complex and multidirectional load requirements of the rings.

New turbochargers have been designed using CF reinforced PAI rotors that are molded in one piece. The rotor permits the use of a smaller drive belt due to the part's lower inertia weight, making for a more compact fit. Heretofore rotors were machined from aluminum billets. By selecting PAI, processing is reduced to 30 seconds. The PAI provides heat stability to 315 °C and approximately 90% of the strength of aluminum. Other engine parts of PAI include composite piston, push rod and connectors.

Use of composites are expected to boost drivetrain efficiency in transmissions. Collaborative efforts between Rogers and Ford Motor Co. resulted in a reinforced phenolic torque-converter reactor that was changed from an axial-pull to a more efficient radial-pull design. The new design, which would be virtually impossible to machine, increases fuel economy and improves acceleration.

11.8 Miscellaneous Uses

Selected applications that incorporate ACM but are more highly specialized and consume low amounts of ACM are as follows.

11.8.1 Medical Uses

(a) Orthopedic and Prosthetic Devices

Both internal and external devices based on ACM can be attached to a patient. Carbon is one of the safest materials for biomedical implants because it is completely inert and compatible with blood, tissue and body fluids. Internal or surgical implant devices are composed of carbon–carbon composites and exhibit greater corrosion and chemical resistance than stainless steel or selective metal alloys. Satisfactory fatigue behavior and toughness, combined with relatively low density of C/C composites make the composite an ideal candidate for prosthetic devices. Replacement of knee and hip joints, dental implants with C/C composites are examples of ACM surgical implants. Heart valves produced with C/C has virtually eliminated wear as a cause of failure. Polymer ligament prothesis is achieved by using fine diameter, high strength UHMWPE (Spectra) fibers with breaking strength above 500 kg and are attached to bone structures.

External systems are artificial limbs for amputees. Strong, light-weight, durable and vibration-dampening characteristics are responsible for rapid displacement of other materials by ACM.

(b) Medical Equipment

X-ray transparency of CF and aramid make these materials suitable for a variety of support systems that are utilized in X-ray medical diagnosis. These include cantilevered support frames, X-ray table tops and CAT scan couches.

11.8.2 Industrial Machinery

Metal machinery parts that undergo reciprocating, oscillating or rotating motions are being displaced by light weight, high strength/modulus, improved fatigue, reduced noise components made of ACM. Higher operating speeds and productivity are achieved with the ACM parts. Examples include parts for robotics, tubes or rollers for film/textile industries and centrifuge rotors.

11.8.3 Racing Cars

A composite chassis made of epoxy/CF can be placed on a commercial auto engine-drive transmission. The Marlboro McLaren Honda racing car (Fig. 19) is an example of ACM providing the elements of speed and crashworthiness to the racing vehicle that has won 7 of 8 championships. Crashworthiness and crash management strategies have been applied in the design of automobiles, particularly racing cars. Maximum energy absorption on impact at high speeds is the goal of the design of the front end of the vehicle for maximum energy absorption to protect or safeguard occupants from forces that cause serious injury or death. Although steel is effective in providing energy absorption (25 kJ/kg), use of orientated CF/epoxy composite results in higher energy absorption of 120 kJ/kg and due to energy absorbed by composite microfracture processes occurring during fragmentation on impact.

11.8.4 Non-Automotive Composite Drive Shafts

Power transmission applications of filament wound composites are rapidly displacing metal drive shafts in non-auto areas. Advantages of the ACM shafts (Fig. 20) versus metal shafts consist of high misalignment tolerance, corrosion resistance (acids and bases), low bending loads on bearings, long single spans, fatigue resistance, little or no thermal expansion, elimination of noise and mechanical vibration. ACM drive shafts are used in high speed, high torque pump and compressor couplings, quill shafts, and cooling tower shafts.

Fig. 19. McLaren Honda racing car

Fig. 20. Cooling tower shaft

11.8.5 Construction

Resin impregnated fibers of S-Glass, aramid or CF are being transformed or pultruded into high strength rods or tendons. These composite tendons are being used for prestressing or post-tensioning concrete for structural applications by providing a high tensile strength, low relaxation under sustained high load, and high modulus of elasticity to limit the extension of the material during stressing. Further advantages include resistance to salt (sea water) and good electrical and magnetic properties. Composite tendons are prime contenders for reinforcement of bridge decks, marine structures, parking garages. The use of composite tendons will require revised civil engineering specifications to include tendons into structures.

Pitch based CF reinforced concrete (CFRC) is succeeding in Japan. Significant improvements in concrete flexural and tensile strengths with increase in ductility and toughness are obtained when 4% fiber is used. A recently completed high rise Tokyo building (32 000 m^2) used 2% fiber and was able to reduce the amount of structural steel from 24 000 tons to 20 000 tons.

11.8.6 Transportation (Rail cars, trains)

Composite train chassis studies are quite active in Germany and France (Fig. 21). A significant advantage that is derived by using ACM is the "built in" dampening effect which eliminates the primary spring of the metal undercarriage. Composite structures can withstand frequent and highly fluctuating load cycles with fewer stress effects than metal frames. ACM weight reduction is an additional benefit.

Fig. 21. Composite train chassis

Recently Alcoa/Goldsworthy Engineering (U.S.) announced that an all composite railway module had been produced for Union Pacific Railroad. It is built from pultruded composite profiles and is believed to be the largest single unit ever assembled from pultruded structural stock (16 m × 2.6 m × 2 m). It was designed and built as an auto hauler for transportation by rail of new autos, and is expected to replace open steel and aluminum containers.

Monorail cars are being fabricated via a monocoque construction technique using an all composite body attached to a steel undercarriage. The composition of the body consists of aramid honeycomb faced with epoxy/glass with some CF in high stress areas such as door frames, window openings and at the juncture of the car body and chassis. These monorail cars are in operation at Disney World, and several U.S. airports.

11.8.7 Masts, Antennas, Radomes

Composite masts, home antennas/satellites, and radomes that are fabricated with ACM are replacing inexpensive glass reinforced materials. The new ACM products exhibit improved strength and dielectric properties.

11.8.8 Musical Instruments

Violins and guitars have been assembled with epoxy/CF sound boards as facings for the instruments. The composite faced instrument gives a high quality sound that is more reliable than wood, remains tuned and provides improved sound radiation qualities when compared to wood.

11.8.9 Non-Ballistic Helmets

High temperature resistant and moderate impact composite helmets for fire protection consist of aramid/glass woven hybrid with vinyl ester resin.

11.8.10 Others

These applications include a portable tri-arch bridge, strategic and tactical missiles, a composite gondola for blimps.

11.9 References

1. "Composite Materials in Aircraft Structures," Edited by D. H. Middleton, Longman Scientific & Technical, Harlow, Essex, England, 1990.
2. D. Nollen, SAMPE 33, 990, 1988.
3. L. Diaz and J. Price, "Proceedings of the Aircraft Interior Materials/Fire Performance," April 4–5, 1989, Wichita State University, Kansas.
4. Temephen D50, Temecon Group International, Inc. Piscataway, N.J. 08854.
5. "Ballistic Materials and Penetration Mechanics," R. C. Laible, Editor, Vol. 5 of Methods and Phenomena: "Their Applications in Science and Technology," Elsevier, Amsterdam, 1980.
6. C. E. Morrison and W. H. Bowyer, Proceedings of the 3rd National Conference on Composite Materials, Paris, France, August 26–9, 1980.
7. S. J. Bless and D. R. Hartman, SAMPE 21, 852 (1989).
8. L. A. Pilato, "Increased Kevlar Composite Ballistics by Matrix-Fiber Interface Study," presented during 21st International Technical Conference of SAMPE, Atlantic City, September 26–9, 1989.
9. C. Blair and J. Zakrzewski, SAMPE Quarterly, 2, April 1991.
10. H. D. Beckmann and H. Oetting, Automotive Eng., May 1985, pp. 34–41.

Subject Index

ablative 87–89
ABPBI 60, 64
ABPBO 60
ABPBT 60
abrasion 8, 9, 65, 69, 109
– resistance 45, 65, 69, 80, 162, 174
acceleration 181
accelerator 14
accuracy 175
acetic anhydride 24
acetylene 30–32
– terminated 11, 30, 31, 66
acid 19, 21, 24
acidic 14
– groups 110
ACM 1–9, 20, 65, 75, 78, 86, 94, 102, 103, 108, 114, 157, 160, 163, 166–172, 174–179, 181–184
– Compounded 3
– Industry 3
acoustic 103, 105, 112
– emission 104, 105
– energy 103
acoustography 105
acrylates 42
acrylonitrile 89, 90
– butadiene 139
activated 43
acylation 48
acyl halide 48
adhesion, fiber-matrix 54, 86, 101, 108, 109, 140
adhesion, improved 110
additives 19, 138,146
AE 104, 105
aerodynamic 174, 175
aerospace 1, 2, 4, 12, 23, 58, 59, 104, 114, 157, 172, 177
– applications 1–4, 12, 58, 145
– markets 7, 23, 114
aft fuselage 53
ageing 46
aggregation 62
Airbus 7, 157, 158
aircraft 1, 2, 8, 20, 35, 46, 86, 93, 101–103, 105, 106, 131, 157–163, 166, 173
–, civil 131, 157, 162, 163
–, composites 157, 159
– design 123, 159
–, fuel efficient 136
– industry 137, 157
– interior 19, 20, 22, 102, 163
– – components 160–163
– –, military 4, 132, 162, 163
– primary structures 66
– structural applications 46, 144
– solvents 53
–, supersonic fighter 131
Airforce Wright Aeronautics Lab 61
airframe companies 136
airplane wing skins 118
AIS-U 105
alcohol 20
alignment 121
aliphatic 14, 22, 29, 62
alkali metal phenoxides 43
alkenyl 38
– ether ketone 38
alloying characteristics 69
alloys 59, 67–69
allyl groups 31
allyl phenyl 38
alternating 34
aluminates 109
aluminum 1, 157, 161, 164, 170, 171, 174–176, 178–180, 184
– chloride 48
ambient temperature 24
amic acid 25, 29, 37, 50
– imide 69
amide 26, 57, 59, 64, 81, 82
amine 24, 34
– curing agent 14, 110
amino-carboxamide 29
ammonia 20
amorphous 34, 43, 45, 46, 51–54, 56, 57, 110, 144, 145, 148, 149
amplitudes 104
Analysis/Testing 97–107

analytical models 128
analyzer axis 86
aniline 36
anisotropic 75, 97, 140
– phase 64
anisotrophy 82, 89
anhydride 14, 24
anhydrous 45
annealing 43
antennas 184
API 23, 24–39, 52
appliances 2
applications
–, end-use 78, 149
–, industrial/commercial 1
–, military 1
–, sports 1
–, structural 144
aprotic solvent 24, 43, 45
aramid 9, 20, 62, 63, 75, 78, 81–86, 93, 94, 99, 108–110, 128, 144, 157, 168, 173, 174, 176, 183,–185
– pulp 63
archery 4, 172, 175
architecture, textile preform 140
areal density 164, 165
armor 86, 93, 94, 164–167, 176, 177
aromatic 12, 14, 15, 24
– diamine 23, 34
– polyesters 58
– rings 57, 58
– substrate 48
arrows 175
aryl
– cyanate esters 40
– diacyl halides 48
– diamine 35
– ether 31, 51, 53
– imide 32
– sulfone 31
aspect ratio, high 61
assembly techniques 148, 149
ASTM 104, 114–116, 121, 122, 129
ATF 46, 53, 157
ATL 142
Atlas 167
atmosphere, inert 90
atomic
– position 112
– resolution 112
– scale 112
ATR FTIR 86, 98, 99
AUSS-V 103
autoclave 57, 104, 148, 150
– cycles 57
autohauler 184
automated 6
automation 144, 148, 160, 179
automotive 5, 6, 20, 58, 104, 179–181
– applications 20, 58, 178–181
avionic 88
avimid 52, 53, 178
AWAC 106
AXAF 169
axial 128, 180
– composite moduli 120
– composite compressive strength 123, 136
– compressive strength 121–123, 131–134
– defects 75, 79
– direction 89, 115, 116
– loading 131
– modulus 88, 89, 117, 128
– properties 117, 127
– tensile strength 121, 122, 127
– thermal conductivity 88, 89
– thermal expansion 88, 89
axis
–, in-plane 115
–, out-of-plane 115
– tension, off 123
aziridines 110

B-2, 4, 106
backbone 47, 56
ballistic 4, 9, 19, 81, 86, 93, 94, 163–167, 177
– applications 163–167
– impact energy 165
– limit 164, 165
ball speed 173
base 19, 21
basicity 14, 43
bats
–, baseball 175
–, table tennis 175
BCB 34
Beech Starship 157, 158
benzocyclobutene 32–34
– imides 33, 34
benzophenone 25
3,3′,4,4′-benzophenone tetracarboxylic acid dianhydride 25
3,3′,4,4′-benzophenone tetracarboxylic acid dianhydride dimethyl ester 26, 29

benzoxazole 33
BF_3 Complexes 14
bicycle 173, 174
bifunctional 109
binary
– compositions 68
– mixtures 68
bismaleimide 11, 32, 34–39, 60, 132, 138, 141, 142, 159, 170
bisoxazoline 21, 22
bisphenol A dicyanates 42, 66
bisphenol A epoxy 11, 12, 14, 139
blades, fan 28
blends 53, 54, 59, 63, 64, 67–69, 101, 149
blimps 185
blister 146
blistering 24
block copolymer 62–64
blood 181
BMAC 53
BMI 11, 16, 23, 34–39, 40, 41, 54, 66, 99, 126, 132, 138, 141, 142, 159, 170, 178
–, brittle 125
–, ductile 125
– modified 141
– toughened 138
boardiness 146
boardy 147
Boeing 157, 158, 162
boiling water 40
bond 23
–, intimate 161
– strength 110, 112
bonding, adhesive 149
–, dual polymer 149
–, fusion 149
Boron 21, 173
bows 175
Bradley turret 164, 166
braid architecture 140
braided 140, 174
braiding 173, 174
brake/clutch pads, facings 106
brakes 23, 89, 174
–, aircraft 23, 162, 163
–, steel 163
branching 48, 56
bridge
–, composite 4
– decks 183
brittle 15, 20, 25, 28, 29, 37, 66, 101, 109, 111, 121, 139
brittleness 28, 32, 35, 37, 66
B stage 20, 26
BT 41
buckling 9, 137
–, non-symmetric 121
– shapes 3D 129
–, symmetric 121
burning, uneven 106
burst
– pressure 85
– strength 85
– testing 170
buses 7
business equipment 2
by-products 13, 20

$CaCl_2$ 82
CAD/CAM 4, 172
CAI 46, 118, 125, 136–142, 145, 151–154
calcium aluminoborosilicate 92
cam elements 179
camshaft 179
canoes 175
carbon–carbon composite 21, 23, 87, 89, 162, 163, 168, 171, 181
carbon fiber 5, 20, 28, 39, 46, 48, 66, 75, 77, 78, 86–93, 99, 108–111, 114, 123, 128, 133, 136, 139–141, 144–147, 159, 163, 167, 168, 170, 172–175, 179, 183, 184
–, chopped 52
– epoxy 168–170, 172, 174, 176, 178, 179, 182, 184
–, high strength/high modulus 157, 172
–, intermediate modulus 117, 125, 126, 132, 134, 136, 142, 145, 149, 155, 159, 175
–, pitch 117, 123
– reinforced 46
– reinforcement 150, 180, 183
–, standard modulus 117, 125, 136, 149, 155
– surface 86, 89, 108, 110
carbonization 89, 90, 163
carbon peak 85, 86
carbonyl 43, 45, 68, 108
carboxylic 38, 108
carboxyl terminated 37
cargo liners 19, 93, 102, 160–162
cars 182
cast 69
castings 105

catalytically 13
catalyst 10, 20, 21, 48, 65
C–C 23, 163, 171
ceilings 161
Celanese test 129
Celazole 59
ceramic 105, 112, 164, 166, 178
ceramic composites 112
cermet 163
chain molecules 79
chain scission 80
char 21–23, 41, 87
char resistance 56
chelates 39
chemical processing 58
chemical resistance 12, 54, 57, 80, 179, 181
chromatography 99
CIFV 4, 6, 166, 167
civil engineering specifications 183
Clean Air Act 6
coagulated 62
coagulation 62
coatings 20, 171
–, protective 160
–, surface 23
co-catalyst 39
co-continuous 39
Cold war 4, 7
commercialized 146
commercialization 81
commingled 140, 146, 147, 149, 150
commodity 1
comonomer(s) 89, 90
compact tension 139
compatibility 59, 67, 68
competitive 1, 6
compimide 16
complex shapes 52
components 27, 145, 164
component assemblies 149
composite
– armor 167, 177
– axial modulus 120
– axial tensile strength 121
– chassis 182
– chassis, train 183, 184
– compressive strength 121, 123, 128–134
– holograms 106
– hull 166
– interphase 108–112
– kites 175
– laminates 145
– laminates properties 155
– matrix resin 149
– mechanical properties 110, 114–119, 144
– modulus 120, 121
– performance 133, 140
– properties 165
– railway module 184
– sample
– shear modulus 120, 121
– skin 131
– strength 120, 121–123, 133, 167
– strength, ultimate 111, 116
–, stressed
– structure, thick 118
– surface 9, 112
– system 120
– tensile strength 121
– toughness 111, 136, 137, 139
– Structure Property Guidelines 119, 120–127
composites
–, damage tolerant 136–142
–, honeycomb
–, polyarylether sulfone/CF 149–155
–, transverse tensile strength 123
–, unidirectional 129
–, rigid
compression 20, 132, 133
– molding 39, 57
Compression Strength
– After Impact 46, 118, 125, 136, 137–142
–, measured 59–61, 129
–, residual 118, 138
– Testing Methods 129–131
compressive
– failure 121, 128
– loading 114, 131, 136, 176
– stress 179
– strength 21, 174, 175
compressive behavior
–, cyclic 137
–, static 137
Compressive Strength/Structure Property Status 128, 129
computer modelling 133
concentration 82
concrete 183
concurrent engineering 8
cone calorimeter 103
configuration 131
conrods 180
consolidation 5, 8, 123, 147–150, 155

constituent property 120, 121
construction 2, 5, 183
consumer products 2
contaminants 26, 92
continuous 9, 18, 21, 65, 66
– filament 146
control
– levers 160
– mode, strain 124
– systems 8
cooling rates, controlled 43, 145
copolymer, ethylene vinyl acetate 111
copolymer(s) 22, 62, 64
copolymerization, addition 38
copper 168
coreactants 13, 14
coreactive 38
cores, crushed 106, 161
corrosion 1
– resistance 2, 176, 181, 182
corrosive chemical 58
cost 1, 4, 6, 10, 43, 170, 177–179
– reduction 179
–, fabrication 160
costs 2, 148, 163, 167
–, maintenance 176
coupling agent, silane 109
coupling agents 109
–, chemical 110
–, titanate 110
covalently bonded 75
CPI 16, 23
crack
– growth 112
– opening force 137
cracks 20
–, cleavage 108
–, matrix fiber 103
CRAG 114
crash-worthiness 182
creep 20, 57, 80
– behavior 43, 47
– resistance 53
critical
– concentration, C_{cr} 64, 82
– length 112
crosslink density 14, 37, 40
crosslinked 10–12, 14, 17, 18–20, 22, 26, 31, 36, 40, 47, 65, 80
– sites 47
crosslinking 11, 13, 20, 35–37, 39, 49, 57, 89, 90, 147
CRT 103
crystalline 34, 43, 45, 50, 57, 58, 75, 80, 85, 91, 108, 144, 146, 148
– boundaries 108
crystallinity 43, 58
crystallinity, transient 50
crystallites 64
crystallographic planes, critical 123
CTBN 139
CTE 91, 108, 163, 169, 170, 178
CuO 44
cure(s) 10, 32, 34, 65, 100, 104, 121, 123, 147, 161, 176, 178
cure cycle 26, 31, 35
cure shrinkage 108
cured systems 12
curing
– agents 13–18, 110
– behavior 20
– reaction 39
cut-outs 114, 119, 125, 133, 134
CV 168
CVD 23, 163
cyanate 16, 38, 99, 111
– esters 11, 39–42
cyanogen halide 40
cycles 178
–, load 183
cycloaliphatic 14
cyclobutene 32–34
cyclodehydrated 25
cyclodehydration 26, 50
cyclophane 29
cyclopentadienyl 25
cylinder 85

damage 102, 104, 106, 109, 134, 160, 178, 180
–, impact 118, 119, 136
–, impact induced 114, 118, 119, 136, 139
–, irreversible 104
–, mechanisms 128
–, reverse side 136
–, through hole 133
–, tolerance 9, 10, 43, 46, 54, 86, 137–141, 155, 162, 176
– zone growth 140
Damage Tolerant Composites 136–142
dampen 4, 172–174
dampening 183
damping peak 67
data base 10, 26, 43, 46, 114, 128, 144, 150, 155, 160

data bases, composite properties 42, 114
DDS 14–18, 40
DEA 100
debonding 104, 111, 112, 121
–, interply 123
–, premature 121
debonds 106
defect 91–93, 97, 101, 103
– populations 90
defects 77, 90, 105, 106, 108, 112, 128
degree of adhesion 141
dehydration, thermal 24
delamination 86, 103, 104, 124, 137
–, on-set 124
delamination zone 136
delaminations, simulated 137
Delta 167, 168
density 84, 85, 93, 163, 178
deply technique 102
design 2–4, 6, 8, 20, 101, 104, 105, 112, 114, 116, 123, 133, 166, 172, 174, 176, 180, 182, 184
–, aircraft 123
– criteria 114, 116, 118, 128
– data, reliable 118
– function 1
–, hardware 114
–, process 146
–, system 1
designers 2, 125, 131, 136, 174, 175, 178
destructive test 102, 103
devices, orthopedic and prosthetic 181
devolatilization 24, 26, 79
diameters 62, 147
diamine 15, 35, 36, 49–51, 54
diamino
– benzene dithiol 93
– diphenyl ether 82
– diphenyl oxide 54
– resorcinol 93
dianhydride 23, 24, 50
diaphram, polymeric 148
diaryl halides 43
dicarboxylated monoanhydrides 24
dichloro benzophenone 45
dicyanate 66
dielectric properties 81, 184
Diels-Alder 25, 33
dienophiles 33, 34
diester diacid 23, 24
difluoroaryl 45
difluoro
– benzophenone 45
– compound 45
difunctional 11, 34
dihalo 43
diisocyanate 24
dilute HCl 40
dimensional 23
– stability 89, 114, 169, 171, 179
dimerization 33
3,3′ dimethyl 4,4′ diamino diphenyl methane 36
dinitro bisimide 53
diphenyl
– ether 48
– isophthalate 59
– sulfone 45
dipotassio dihydroxy benzophenone 45
diradical 29
disbonding, fiber/matrix 129
discontinuous 66
– phase 111
disodio
– bisphenol A 44, 53
– bisphenate 45
dispersion 20, 64
dispersed 18, 61, 65, 66
dissimilar
– phases 9, 65
– polymers 65
DMA 11, 38, 60, 100
DMSO 44, 45
domain size 39
double cantilever beam 138
– specimen 137
– tests 137
double diaphragm 148
drape 10, 20, 26, 34, 67, 138, 142, 146, 147, 161
drapeability 149
drawing ratios 77, 79, 80
DRIFT 98, 99
drive shafts 6
–, automotive 179
–, composite 180, 182
drivetrain 180
DSC 30, 33, 41, 47, 79, 81, 100, 101
DSM 79
DTA 100
ductile 17, 18, 101, 111, 121
– phase 28, 38
– resin 37, 39
ductibility 123
ductility 40, 53, 57, 65, 183
durable 4, 101, 144, 160, 172, 177, 181
durability 9, 171, 174, 175, 178

Durimid 50
dyes 105

economics 8
edge
– damage 121
– delamination 137, 151–154
– delamination strength 120, 122, 123–127
– delamination tension 138
–, delamination, visible 122
– replication 106
EDS 120, 122, 123–127, 137, 151–154
E Glass 5, 20, 21, 92, 93, 139, 157, 161, 164, 175, 176, 179, 184, 185
eight membered ring 33
elastomeric
– additives 139
– modifiers 137, 139
elastomers 15, 17, 139
electrical 2, 80, 81
– applications 20
– conductivity 88, 91, 168
– insulation 81
– laminates 41
– properties 80, 81, 183
– resistance measurements 112
electron diffraction 89
electronic(s) 2, 58, 59
– applications 59
– equipment, video 106
– shearography 105
electrocyclic 32
electrophilic 43, 48
electropolymerization 110
electrostatic 146
elemental analyses 90
elevated temperature 10, 20, 22, 25, 29, 40, 45, 118, 119, 123, 126, 127, 131, 136, 140, 144, 155
elongation 63, 66, 84, 92, 101, 128
ELV 167
elevators 157
empirical 127, 129
end
– crushing 131, 137
– groups 25, 29, 30, 31, 34
– load 137
– tabbing 124
– tabs 126
– users 133
end capping 24
– agents 48
energy 22, 111, 138, 163, 165, 173, 178
–, absorb 9
–, absorption 182
–, transmitted 173
engine
–, composite 20
–, core 168
– duct 28, 159
–, internal combustion 179
– oil 179
– turbines 23
entangled 65, 79
entanglement 62, 79
–, molecular 62
entrapped 65
environment 9, 22, 136
–, harsh 118, 119, 136
–, humid 112
environmental
– degradation 108
– factors 101
– performance 101, 120, 126
– properties 101
– resistance 10, 53, 75, 97
environments
–, in service 131
–, solvent 120
epichlorohydrin 11
epoxy 11–18, 29, 34, 38–40, 63, 66, 99, 100, 109, 110, 112, 117, 118, 126, 128, 132, 133, 134, 138, 139, 142, 160, 168, 175, 184
–, brittle 111, 125
–, brominated 41
–, damage tolerant 118, 126, 132, 139
–, ductile 125
–, flexible 62, 111
–, glass 21
– modified 17, 139, 173, 175
–, tough 18
–, toughened 140, 170
equations 116, 120
ESCA 85, 97
ESPI 105, 106
ether 22
– amide copolymer 21
– equivalences 56
– linkages 43
eutectic 29
exit cone 168
expendable systems 167
exponent 82
extensometers 121
external angle plies 124

extraction 79
extrudate 79
extruded 79
extrusion, solid state 79

F-22, 4, 5
FA 102, 103
FAA 102, 103, 157, 160, 162, 176
fabric 146, 147
– comingled 149, 155
fabrics, woven 20, 146, 148, 179
fabrication 2, 3, 9, 10, 42, 43, 53, 57, 140, 142, 145, 146, 148, 166, 168, 173, 176, 177, 179
face sheets 131
failure 181
– analysis 102
–, brittle 101
–, buckling 129
–, catastrophic 8
–, composite 128, 133
–, compressive 128, 129, 131
–, interfacial shear 123
–, matrix shear 123
– mechanisms 89, 123
failure mode
–, premature 131, 134
–, splitting 121
–, brittle longitudinal 121
–, composite axial compression 123
–, composite compressive 123
– modes 89, 101, 104, 121, 123–125, 128, 129, 133, 134
–, kinking 129
failure strains
–, higher 129
–, low 129
fan blades 28
fastener pullout, resistance 162
fasteners 2, 149
fastening 133
fatigue 97, 101, 102, 140, 172, 175, 180, 181
– performance 43
– resistance 1, 182
– strength 20
– stress 20
F–C 48, 49
fiber 23, 97–99, 147, 165
– adhesion 39, 67
– alignment 121
– axial modulus 136
– axis 90, 115
– buckling mode 78
– bundles 92, 147
– characteristics 141
– compressive strength 78, 123, 129
– cross section 93
– damage 140
– diameter 77
– drawing 79–81, 92
–, experimental 75
– failure 123
–, gel 79
– impregnation 10
– impregnation, degree 145
–, interior 64, 93, 97
–, low modulus 128
– matrix 23, 109, 111, 112, 124, 129, 140
– manufacturers 3, 86, 97
– microbuckling 104
– misorientation 78
– misalignment 121
– modulus 77, 120
– molecules 77
– phase 108
–, pitch 88, 89, 128
–, pre-impregnated 3, 145
– pullout 104
– reinforcement 123
–, reinforcing 3, 9, 61
–, resin ratio 103
– shear treatment 122, 127
–, short 159
– spinning methods 78, 91, 93
–, spun 64
–, stiffer 128
– stiffness 104
– strength 75, 77–78, 86, 121, 165, 167
– surface 85, 93, 97, 108–110
– technologies 140
– tensile strength 141
– volume 28, 120, 141
– volume fraction 120, 121, 140
– waviness 129
– wet out 140, 146, 147
– functionality 141
– treatment 78
fibers, organic 78, 93, 94, 108
fibers orientation 148
fiberglass reinforced plastics 1, 2
fibrillar 78
fibrillated 63
filament 147, 148
– winding 3, 6, 7, 39, 67, 85, 93, 104, 111, 147, 167, 168, 182

filler 19
film
– adhesive 139
–, gold 112
–, thermoplastic 147
film(s) 50, 58, 59, 69
films, clear 59
fire 102, 160, 166, 176
– detection 162
– protection 185
fishing rods 4, 172, 175
flame 18, 22, 56
– resistance 56
– retardancy 41
flammable 22, 160
flammability
– resistance 176
– standards 176
– testing 102
flaws 8, 97, 101, 105, 123, 142
flexible 26, 58, 62, 64, 82, 111, 164
– coil 62, 63
– linkage 50
flexibility 4, 63, 172
–, formulation 20
–, chain 43, 53
flex point 172
flexural strength 21
flight experience 160
flow 53, 146, 147, 166
flowing air 52
fluorinated 28, 54
fluoro 54
folds 79
foreign inclusion 108
formaldehyde 19
formed details 149
forming 148
fracture 9
– energy 15–17, 37, 38, 137
– inducing flaws 108
– mechanics 140
– surface area 39
fracture toughness 9, 17, 29, 37–39, 51, 66, 67
–, interlaminar 10, 124
fragmentation 22, 182
frames, door 184
Friedel-Crafts 43, 48, 49, 69
friction 162
– coefficient 163
fringe lines 106
FST 19, 20, 102
FTIR 41, 68, 89, 100, 109
–, photoacoustic 86
fuel
– blends 6, 7
– costs 136, 157
– economy 181
fuels, automotive 6
functional groups 108, 110
functionality 11, 14, 15, 51, 108
fundamental properties 116
fuselage 157, 159
–, fighter forward 148

gas
– burner 103
– mileage 179
gasoline 7
gel
– sol 80
– spinning 75, 79
gels 79
GEO 168
glass
– beads 111
– fiber 85, 92, 93, 108, 109, 111, 112
– size 109
glycidyl 12, 14, 15
golf 4, 172
gondola 185
GPC 99, 100
graded modulus interphase 111
graft copolymer 62, 63
graphite 86–92, 111, 178
– molecular sheets 75
graphitization 22, 90
gravity, center of 174
GRO 169
guidelines, composite structure-property 120–127
guitars 184

halogenated compounds 105
hardening agents 10
hardware 144, 167, 171
HDT 16, 58
heart valves 181
heat
– capacity 56, 163
– resistance 9, 11, 14, 22, 38, 39, 42, 48, 53, 65, 66, 163
– resistant 15, 66
– shields 23
– treatment 92

heat release 22, 160, 161
– –, peak 160
– – tests 102, 103, 160
– –, total 102
helicopter 2, 118, 157, 166
helmet 4, 86, 94, 164
helmets, non-ballistic 185
HERTIS 105
heterocyclic 12, 14, 58–61
hexa 19, 20
hexafluoro isopropylidene 51
– bisphthalic acid 28
high
– modulus 17, 51, 85
– performance 9, 13, 18, 20, 23, 26, 39, 41, 42, 50, 57, 65, 66, 68, 85, 93, 97
high pressure 7
high solids 24
high stress 184
high strength 1, 9, 42, 45, 57, 58, 66, 75, 81, 174, 181
high technology 4, 174, 175
high tenacity 79
high tensile strength 183
High Performance Fibers 75–94
high temperature 26, 28, 29, 36, 39, 41, 42, 46, 50, 51, 57
– applications 23, 28
hole 118, 119, 125, 133, 134, 136, 155
holes
–, circular 137
–, machined 119, 133
holography 105, 106
homopolymerize 65
honeycomb 20, 81, 105, 106, 171, 184
– composites 106
– core 131, 160
– defects 105, 106
hot
– drawing 79
– melt 22, 35, 39, 146, 147
hot/wet
– behavior 110, 138
– compressive testing 136
– conditions 126
– environment 14, 15
– performance 9, 18
– resistance 14, 15
– temperature 12
HPLC 27, 99
HST 169
HUMMER 164
hybrid 42, 86, 146, 167, 173–175, 185
hydraulic fluids 53
hydro-forming 148
hydrogen 168, 170
– bonded 75, 82
– bonding 82
hydrolytic 45
hydrolytically 24
hydrophobic 40
hydroquinone 45
p-hydroxy benzoic acid 58
hydroxyethyl methacrylate 38
hydroxyl 40, 108, 110
hydroxy naphthanoic acid 58

IFSS 111, 112
ignition 102
IITR 129
ILSS 110
image
– areas, dark 103
–, dark 103
–, electronic 106
–, hologram 106
–, storage system 106
images
–, 3D 112
–, dark field 64
–, stressed 106
–, unstressed 106
–, visual 103, 105
imidazoles 39
imide 23, 25, 26, 32, 68, 69, 99, 100
imidized 28
impact 102, 137, 138, 159, 173, 175, 176, 182, 185
– energy 165
– event 136
–, izod 139
–, low velocity 137, 139
– modification 37
– resistance 139, 140
– shear, post 140
– tests 162
impedance 86
imperfections 79, 103
implants
–, dental 181
–, surgical 181
impregnation 23, 146, 147
impurities 27, 108
inclusions 91, 104, 129
incompatible, moderately 65
industrial 144
industrial machinery 181

inert 45, 58, 90, 181
inertness 110
inhomogeneities 77, 106
injection moldable 58, 65, 68
injection molded 20
inner cowl 28
inorganic 21
inserts 105
in service 97, 101
– molding 123
in situ 26, 62
– polymerization 62
instruments 169, 184
insulation 167
insulation, rubber 106
interchange 69
interdigitated 60
interface 61, 89, 97, 172
–, fiber/resin 122, 124, 141
–, matrix/fiber 104, 114
–, interphase 120
– requirements 46
interfaces, weak 112
interfacial
– adhesion 85, 110, 112
– properties 117
– region 108
– testing system 111
interfacial bond 108, 121
–, weak 121
– strength 111, 112
interfacial shear bond strength 129
interfacial shear failure 121
interfacial shear strength 122–124, 127
interfacial strength 85, 112, 121
–, fiber/matrix 125
interference 106
interferrometric 105, 106
interferrometry 106
interiors, marine vessel 102
interior, submarine 176
interfilament contact 147
interlaminar shear 28, 123
– strength 122, 123
interlayer shell 111
interleaf
–, adhesive 139
– concept 139, 140
interleafing 139
intermediate 27, 100
– material 145
– strength 23
intermolecular reactions 59
Interpenetrating Networks 65–67
interphase 108–112
–, fiber/matrix 108, 112
–, flexible 111
–, graded modulus 111
–, rigid 111
interply region 124
intractable 24, 54, 90
intramolecular shear strength 122, 123
IPN 18, 29, 42, 60, 65–67, 101
irradiation 80
–, electron beam 80
isoimide 31
isopropylidene 43
isotactic 80
isothermal 10
– ageing 28
– contour lines 106
isotropic 64, 82, 97
– liquids 57
ISS 98

Jeffamine 29
jet engine 35, 52
jet-wet spun 75, 82, 93
joining techniques 176
joint ventures 3, 8
joints 2
–, knee and hip 181

Kadel 44
kayak, paddles 175
K_{ic} 18
ketimine 46, 49
Kevlar 19, 57, 61, 63, 75, 76, 78, 81–86, 99, 161, 162, 164–167
kinetic theory of fracture 77
kinetics, cure 110
kink band formation 128
kinding 128, 129
kinks 79

lactone 110
lamellar 79, 80
lamina 104, 114, 116, 117
–, higher stiffness 136
– orthotropic 114
– properties 114–119
– tensile strengths 117
–, unidirectional 137
laminate
– composite, unidirectional 139

– configurations 118
– mechanical properties 110, 114, 118, 119
– orthotropic 139
– penetration 136
– shear, undirectional 139
– tensile test 123
laminated plate theory 116
LARC-CPI 44, 51
LARC-TPI 44, 50, 51, 59, 68
LRAC-13 29
LARC-160 29
LARC-RP40 29, 66
laser 22, 77, 106
– power 106
laser beam, focused 105
–, split 106
Laser Raman Spectroscopy 109
lateral
– interactions 75
– packing 75, 79
– support 129, 137
launch 167
– systems 167, 168
– vehicles 168
LCP 57–61
leaving group, halide 43
LEO 168
Lewis Acid 48
Lewis Base 48
licensee 50
licensing 50, 79
life
– cycle 144, 163
–, service 162
lighter 175
light weight 4, 20, 23, 136, 162, 176, 181
– panels 81
linear chain 65
linear polymer 56, 66
liquid 19, 20, 23
– crystallinity 75
– penetrating agents 105
liquid crystal 57, 58, 109
– display 105
liquid crystalline 57, 60, 82, 93
– polymers 145
load
– bearing 93
–, compressive end 134
– direction 133
– distribution 9
–, full 180
–, high 183
loading, in service 123
loads, bolt 179
longitudinal
– direction 91, 114, 155
– splitting 121, 128, 129
longitudinal composite performance 134
longitudinal composite strength, low 139, 140
Los Alamos National Labs 61
low
– density 75, 87, 92, 93, 181
– flow 139
– MW 10, 24, 26, 39, 42, 49, 54, 99
– temperature 136
– viscosity 24, 28, 42
lubricity 109
lyotropic 57, 58

mach 160
machinery 148
–, automated tape lay-up 148
macromechanics 117, 119
macromolecule 10, 57
macroscopic
– composite 63
– propagation 104
magnesium aluminosilicate 92
magnetic properties 183
maintenance 161
maleic anhydride 34, 37
maleimide 25, 33, 34, 36, 37
mandrel 171
manual spot welding 148
manufacturing 43, 148, 167, 171, 179
– costs 148, 155
marine 1, 2, 102, 176, 183
market, end-use 3
market segments 1, 3
markets, specialized 8
Mark-Houwink 82
mass transportation 5
masts 184
material
– scientists 101
– systems 148
materials suppliers industry 114, 136
matrices
–, crystalline 145
–, damage tolerant 123
–, high temperature 144
–, toughened 141

matrix 23, 109
– ductility 121
– phase 108
– strength 139
– toughness 140
– ultimate strain 123
matrix modulus 120, 121, 123, 139
–, low 121
matrix resin 2, 3, 8, 9, 42, 48, 85, 99, 173, 176, 178
–, yield strength 129
matrix shear
– modulus 129
– strength 124
MDA 25, 34, 35, 54
mechanical 32
– integrity 47
– mixing 67
– properties 17, 18, 31, 32, 38, 40, 41, 44, 48, 51, 60, 62, 64, 67, 89, 90, 97, 101, 144, 168
– properties, maximum 10, 21
– strength 23, 46, 57
mechanics, continuum or solid 116
medical uses 181
melt
– blending 69
– impregnation 48
– processable 29, 58
– processability 53
– stability 48
– viscosity 10, 39, 48, 50, 146
metal 1, 4, 20, 102, 105, 157, 161, 163, 166–176, 178, 179, 180, 181, 182, 183
– carboxylates 39
– fatigue 101
– skins 131
metallic 148
metering truss 169
methane sulfonic acid 93
methanol 26, 28
methanolic 26, 28
methods, assembly 148
methylene chloride 48
methylene diisocyanate 53
methyl pyrrolidone 82
micellar 80
microanalytical 89
microbuckling 104, 129
microcracking 9, 27–29, 35, 37, 66, 67, 104, 121, 176, 178
– resistance 27, 28, 66, 110, 141, 168, 170, 171
microcracks 27, 112
microfailures 101
microfibrillar 78
microfibrils 62
Microfocus Unit 105
microfracture 182
micromechanics 116, 117, 119, 120, 127, 128
microphase 17
microporosity 108
microstructure 68, 89, 123
microtexture 90
military 1, 2, 4, 7, 8, 23, 50, 59, 94, 157
– application 50, 59
– programs 43
millibar 106
misalignment 105, 182
miscible blends 59, 149
missile(s) 28, 86, 87, 167, 185
modelling, thermal mechanical 148
modified 23
modifiers 17, 108
–, surface 109
modulus 9, 18, 20, 126, 132, 133, 140
– of elasticity 183
– retention 126
moisture 54, 80, 87, 97
– absorption 14
– absorptivity 87
– gain 16
– regain 82–85
– resistant 40
– sorption 54
moldable material 54
molded 179
molding material 19, 20
molecular
– alignment 58
– architecture 57
– composites 33, 34, 61–65
– entanglement 62, 65
– level 64
– orientation 75, 79
– weight 82, 139
– weight distribution 79, 80, 100
molecularly dispersed 61
molten state 68
monitoring 100
monoanhydrides 24
monocque 174, 184
monomer(s) 13, 24, 26, 28, 37, 39, 46, 65
monorail cars 184
morphology 37, 64, 66, 79, 82, 85, 90, 93, 100, 110, 138, 145

motor cases 35, 93
MPDA 63
MPD-I 81
MRI 112
Mn 25, 26, 29, 62
multi-component 17, 35
multidirectional 99, 180
multifilament 90
multifunctionality 12, 14, 18, 40
multi-phase 42
multiple bond 59
multi-step 26, 35, 40 , 89, 163
musical instruments 184
Mw 82
MW 20, 22, 24, 25, 28, 39, 54, 80, 99
MWD 79, 80, 100

nacelle 28, 159
nadic 24–32
– acid 29
– anhydride 25, 29
– ester 28
NASA 4, 24, 26, 29, 50, 66, 133, 136, 137, 139, 167, 169
NASP 160
natural gas 7
naval vessels 176, 177
NDT 97, 103–106, 121
neat 144
network 10, 11, 13, 26, 36, 39, 65, 80
– interstices 65
nickel 178
NIST 102, 176
nitrile 38
nitrogen 45, 85
– peak 86
nitro groups 47
NMP 45
NMR 98, 100
–, solid state 22, 49, 90, 100
–, CP/MAS 22, 85, 99
noise 167, 179, 181, 182
Nomex 20, 57, 81
non-bonded 75
non-combustible 59
non-crystalline regions 80, 99
non-invasive 97, 102
non-linear 132, 133
non-linear behavior 78
non-load bearing 86, 160
nonyl phenol 39
norbornene 24–30
5-norbornene-2,3-dicarboxylic dianhydride 25
5-norbornene-2,3-dicarboxylic acid monomethyl ester 26
norbornenyl 26
Northrup test 129
–, tabbed 129
Noryl 68
nose cone 35
notches 114, 118
novolak 11, 15, 19–21
nozzle 87, 88, 168
NR-150 B2 44, 51–53
nucleating effects 110
nucleation 108
– densities 110
nucleophilic 24, 34, 43, 44, 47, 53
nucleophilicity 43
nylon 3, 62, 63, 66, 81, 164, 165

oars 175
objects, low mass 169
ocean submersibles 179
OEM 3
OHC 118, 128, 133, 134, 136, 138, 140 151–155
OHT 118, 138, 140, 151–154
oil
– additives 179
– burner test 162
oligomer 10, 11, 13, 25, 26, 29, 30, 32, 36, 39, 65, 99, 100, 108
oligomerization 39
oligomer systems 65
opaque 105
optical 105
– microscope 111
orbital travel 171
orbits 168–170
ordered
– molecules 57–61
– polymers 58
ordnance 105
organic solvents 32, 36
organosilanes 109
orientated 58, 78
– crystallization 79
orientation 64, 75, 77, 90, 91, 97, 141, 172
orthopedics 181
oscillating 181
OSU 22, 102, 103, 160
outer angle plies 124

outlife 145, 155
out-time 10, 142
oxazoline ring 41
oxidation 22, 90
– resistance 43, 90, 91, 171
–, surface 46, 110
oxidative stability 31, 41, 87
oxidizing agents, liquid and gaseous 110
oxy-dianiline 54
oxygen 22, 85, 168, 170, 171
oxygen consumption 103

PAA 24
PAE 43–49
PAEK 36–38, 53
PAI 54, 145
PAN/CF 76, 80, 86–92, 123, 128, 145, 163
panels
–, ceiling 160, 162
–, instrument 160
–, interior 160, 161
–, sandwich 161
paracyclophane 29, 30
paraffin oil 79
parameters 149, 150
particles 105
partition 102, 160
parts 2, 42, 167, 171, 178, 181
–, engine 180
– inspection techniques 160
–, large 148
–, molded 59, 160
–, non-contoured interior 161
PAS 98
payload 167–170, 176
PBI 59, 60
PBO 60, 61, 75, 76, 78, 82, 93, 108
PBOX 16, 21, 22
PBT 60–62, 64, 75, 76, 78, 82, 93, 108
PEEK 44–48, 109, 110, 145, 155, 159, 168
PEI 16, 39, 41, 44, 48, 53, 54, 66, 68, 69, 109, 145, 155
PEK 44, 45, 48, 49, 69, 145
PEKK 44, 48
Pentagon 4
performance 1
personal protection 164
PES 44
PF 19–22
phase
– inverted 66
–, second 15
–, separate 18
– separated 39
– separation 39, 62, 64, 65, 68
phenol 19, 21, 22, 37
phenolics 11, 18–23, 26, 40, 41, 87, 99, 161, 163, 164, 168, 179, 180
–, modified 161, 162, 164
phenoxide 43
p-phenoxy benzophenone 48
phenylene bisoxazolines 21
phenylene diamine 28, 54
p/m phenylene diamine 51
p-phenyl biphenol 58
photoacoustic 86
phthalimide 31
PI 2080 44, 53, 54, 59, 68
PIC 118, 125, 136–142
piezoelectronic sensors 104
pipes 104
piston rings 180
pitch 23, 163, 183
– mesophase 86–92
– petroleum 88
Pitch CF 76
pitch fiber, high modulus 129, 145, 168, 171
–, ultra high modulus 145, 171
plasma(s) 57, 110
plasma treatment 110
plasticizing 14, 109
plies 122
–, number 133
–, stacking alternate 147
ply
– orientation 133, 137, 139
– stack-ups 148
– thickness 140, 141
PMR-15 16, 26–29, 41, 52, 66, 159, 178
PMR-30 28
PMR-60 28
Poisson's ratio 120, 124
poles 174, 175
–, vaulting 175
Polimotor 20, 179
polyarylonitrile 79, 80, 86, 89
polyamic 24, 26, 100
polyamide-imide 53, 54, 69, 144, 145, 180
polyamides 57, 68, 81, 110
polyamines 14, 36
polyaminoamides 14
polyaryl

– sulfide 44
– sulfone 17, 18, 43, 145, 149
polyarylene sulfide 145
polyaryl ether 43–49, 66, 68, 144, 145, 149, 155
– ketones 37, 38, 69
polybenzimidazole 59, 66, 68
polybenzothiazole 60
polybenzoxazole 60
polycarbonate 42, 49, 66, 111
polycyclic 22, 40
polyester 11, 58, 110, 161, 164, 166, 176
polyethylene 79
–, UHMW 9, 75, 76, 78–81, 93, 94, 109, 110, 164, 181
– oxide 62
polyfunctional 12, 14, 35, 40
polyhydantoin 16, 18, 39, 66
polyimide sulfone 66, 67
polyimides 11, 16, 23–39, 53, 56, 66, 68, 145, 159, 170
poly-L-glutamic acid 62
polymer
– blends 51
– compatibility 59
– concentration 80
– miscibility 59
– volume 9
polymeric
– materials 128
– thiols 14
polymers
–, addition 78
–, amorphous 149
–, ductile 128
–, natural 109
–, softer 128
–, synthetic 109
polyphenols 14
polyphenyl quinoxaline 52
polyphosphoric acid 93
polypropylene 79, 80
polyquinazolines 52, 56
polyquinoline 56
polyvinyl
– acetate 109
– alcohol 79, 80, 109, 110
– fluoride 161
poly-p-xylylene 29
porosity 103, 112
–, interply 105
porous 25
post
– crash 102
– cure 20, 54
– cured 20
– curing 10, 11
– impact condtions 155
– surface treatment 57
Post Impact Compression 118, 125
powder 20, 50, 56, 59, 146, 147
–, white 124
power 172, 173
PPD-T 62, 63, 78, 81–83
PPS 44, 46, 56–57, 109, 110, 112, 145, 159, 179
PPSS 44
precipitation 24
precursor 33, 86
preform 146, 147
–, net shape 140
premium 59
prepolymer 25, 26, 65
prepreg 3, 9, 27, 28, 39, 43, 50, 53, 57, 62, 67, 97, 99–101, 103, 116, 138, 142, 145, 148–150, 155, 161, 168, 171, 174, 175
– morphology 141
– process 146
–, damage tolerant 142
– stacking sequency 105
– tape 146
–, thermoplastic 145–148
–, tow 146, 168
prepregging 62
press forming 148
pressure 93, 106, 161
– hull 176
– tanks 104
– temperature cycle 147
processable 24, 29, 31, 39, 49
processing 8–10, 23, 28, 43, 53, 65, 146, 147, 150–155, 157
–, diaphram 148
production
–, cost effective 148
– rates 6
programs 7
projectiles 86, 164, 165
propellant, liquid or solid 167, 168
properties 9, 111
pseudothermoplastic 145
PSF 18, 36, 38, 39, 41, 44, 45, 66, 144
pultruded 175, 183, 184
pultrusion 21, 148
pyrolysis 22, 23

Quadrap 146, 149, 150
quality control/assurance 10, 27, 28, 43 97, 148
quantum chemical calculations 56
quasi-isotropic 133, 137, 140, 141
– compression strength 151–154
– tension strength 151–154

racing cars 182
Radel 44, 145, 149–155, 159
radar 167
– profile 157
– transparency 81
radial pull 180
radiography 106
radome 81, 184
radomes, aircraft 106, 160
RAE 129
rail cars 183
Raman
– microprobe 89
– Spectroscopy 98, 99
random coil 61, 62
raw materials 3, 4, 6, 8, 97
rayon 86–92
rays
–, gamma 105
–, neutron 105
–, X-ray 105
reactive
– diluent 22
– monomer 65, 108
– oligomers 39
reciprocating 181
recombination 22
recoverable 167
reduction 82
reinforced fiber preforms, 3D 140
reinforcement
– efficiency 62
– geometry 120
reinforcing agent 9, 26, 61, 75, 78, 94, 97, 104
relative humidity 126
reliability 87, 167
repeat unit 58, 67
repair 57
repairability 144, 160
requirements, end-use 117, 131
residual properties 118
resin gelation 140
resins, neat 139, 144
resole 19, 20–23
reverse Diels-Alder 25
rheological 11, 50, 89
risk 160
rigid 10, 35, 40, 57, 81, 82
rigid rod
– molecules 57, 62, 82
– polymers 58–62, 93, 94
rigidity 11, 20, 53, 65, 174
rigidized 62, 82
ring closure 22, 24
ring opening 22, 32
rocket 87, 88, 167
– engines 106
– nozzle 87
rocket motor
– cases 3
–, solid 93, 105, 168
rod
– diameter 123
– torsion 123, 175
rodlike 57, 82
rods 183
–, exercise 175
roll forming 148
rotating 181
rotation 81
rotor 180
–, centrifuge 181
roving, fiberglass 90
RS 98
RTM 21, 22, 140
rubber 15, 17, 37, 38, 106, 148
rudder 157

SACMA 114
safer 4, 172
safety 162
sailboards 175
sandwich beam
– sample 131
– test 129
SANS 68, 98, 101
satellites 168, 184
–, commerical and military 168
SAXS 98, 101
scans, A, B, C 103
scanning angle 86
screw-driven 124
SEC 99
SEM 38, 64, 89, 98, 101
semi-crystalline 46, 51, 54, 56, 57, 110

semi-IPN 18, 29, 38, 39, 42, 65–67
semi-rigid 56
sensitivity 126
sensors 8
sequence
– requirements 46
– units 100
servo-hydraulic test machine 124
S/S-2 glass 19, 75, 76, 78, 92–94, 109, 144, 161, 162, 164–168, 174–177, 183
shear
– crippling 128
– modulus 9, 36, 121, 129
– modulus retention 126
– properties 123, 127
– strength 54, 117, 123
shelf life 10
ships 176
shish kabob 79
shock
– dampening 174
– loading 102, 176
shrinkage 22, 35
shortbeam
– shear 123
– shear strength 37
short excursions 59
shuttle 168, 170
– orbiter 28
sidewall, deck house 176
sidewalls 160, 161
signals, attenuated 103
SIMA 90, 98
single phase 59, 67, 68
sintering process 59
SIRTF 169
size 179
sizes 28
sizing 90
– agents 28, 109
skin-core effect 93, 108
skins 118, 131
–, wing 136
skis 174, 175
SLAM 105
sleds 175
smart/intelligent structures 8
smart systems 8
SMC 21
smoke 18, 24, 102
– emissions 161
– release test 102
– toxicity 102
sodium 87
– hydroxide, 40% 40
– hydride/DMSO 110
solar arrays 168, 169
solid
– crystals 57
– fuel 106, 168
– mechanics 116
– state 22, 79
solution 146
solvent
– free 29, 57, 149
–, residual 146
– resistance 10, 11, 43–46, 53, 54, 57, 59, 69
– resistant 37, 49
– sensitivity 43
solvents 20, 24, 58, 127, 146, 147
sound 185
–, sonics 103–105
– waves 103
Soviet Republic 7
space
– applications 50, 167–171
– hardware 86, 88
– modules, self-contained 167
– program 4
– shuttle 168, 170
– station 168
– vehicles 23
spacecraft 169
spall 164
span-to-depth ratio 123
specific
– heat 178
– modulus 75, 77, 80, 81. 85, 93, 94, 131
– parts 147
– strain 131, 132
– strength 1, 61, 75, 77, 80, 81, 85, 93, 94, 131, 132
speckle 106
Spectra 61, 75, 78–81, 128, 164, 165, 167, 173, 174, 181
spectroscopy, electron energy loss 89
speed 81, 174, 176, 182
–, rotating 166
–, winding 81
speeds, operating 181
SPI 104
spinline stretching 79
spinning speed 79
spinodal morphology 66
spoilers 157
sporting goods 86, 114

sports
– equipment 8, 171, 172, 175, 176
– hardware 4, 172
Sports/Leisure 171–176
spun, wet 90
sputtered 112
SRI 60, 61
SSLCP 60, 61
stability 10, 21, 57
strach 109
stealth 4
– bomber 106
steel 1, 163–165, 174, 176, 178, 179, 181–184
sticks, hockey and lacrosse 175
stiff 26
stiffness 1, 52, 57, 58, 75, 81, 85, 89, 93, 94, 106, 114, 138, 168, 172–175
–, tortional 172, 174
stitching 140
STM 98, 112
stoichiometric 14
storage 27
–, limited 146
– stability 145
stowage bins 102, 160
strain
– concentrations, inherent 123
– gauges 121
– increment 133
–, ultimate 121, 131
–, ultimate matrix 121
strains at failure, ultimate 116
strength
–, flexural 183
–, peel 161
–, yield 129
stress 65, 102, 108, 111, 112, 175
– concentrations 9, 111, 114, 139
– levels 131
– risers 118, 133, 134, 136, 155
– strain curve 132, 133
– transfer 108
structural
– applications 183
– integrity 46, 87
– performance 123, 140
– properties 10
– property relationships 13, 114, 116, 120, 124, 134, 136, 140–141
styrene 38, 42
– acrylonitrile resin 111
submarines 176
subzero 101
sulfolane 45
sulfuric acid 82, 83
surface
– analysis 85
– area 175
– effects 109, 110
– energy 109
– functionality 140
– topography 106
– treatment 90, 93, 109, 141
superior performance 1, 4
survivability 176
swelling 9
– agent 48
swells 65
swing/strike 172
swirl frame 28

tabs 105, 129
tack 10, 20, 26, 34, 57, 67, 138, 142, 146, 147, 161
tack-free 20, 146, 147
tackifiers 146
tailor 2
tank(s) 7, 164, 168, 170
tape 116, 146
–, interlaced 149, 150, 155
–, unidirectional 146, 147
– winding 148
tapes, slit 146, 147
TBA 67
Technora 76, 82, 84, 85,
telescope 169
television camera 106
TEM 64, 89, 98, 101
temperature
–, controlled 126
– rise 103
–, serivce 149
–, skin 159
–, use 139
–, processing 144
tendons 183
tennis rackets 4, 172, 173
tensile
– loads 136
– modulus 51, 79, 80, 87, 91, 121, 139
– side 131
– strain, ultimate 139
– strength 21, 48, 51, 60, 78, 79, 80, 85, 87, 90, 91, 139, 183
tension 132, 133
–, +/−45 120, 126, 127, 151–155

terephthalic acid 58, 93
terephthaldehyde 35, 36
terminus 36
ternary 39
tetrtiary amines 14
T.E.S.T. 103
test
– coupons 121, 123, 124, 126, 131, 133
– fixtures 129, 134, 137
– methods 78, 114, 125, 136
– orientation 140
– techniques 121, 129, 134, 140, 141
testing 97–106
–, drop weight 137
tests, standardized 137
tetraamino biphenyl 59
tetracarboxylic acid dianhydride 23
Texaco Test 129
textile preform architecture 140
TGA 30, 100
TGMDA 12, 15, 18, 20
Tg 10–12, 14, 17, 29, 32, 33, 36, 38, 41, 46, 48, 52, 54, 56, 66, 100, 138, 139, 144, 145, 149, 159
– correlation 56
–, cure 32
–, separate 69
– sharp 69
–, single 59, 69
–, ultimate 54
theoretical 80, 127
– models 111
thermal
– analysis 99
– conductivity 87, 89, 91, 163, 168, 171, 178
– cycling 28, 171
– erosion 87
– expansion 1, 89, 163, 182
– mass 178
– resistance 12
– shock resistance 162
– signatures 167
– stability 28, 31, 45, 87, 93, 94
– stress 22
thermal strains 123
–, residual 121
thermally stable 40, 43, 90
Thermid 16, 30, 31
thermoacoustic technique 112
thermography 106
thermooxidative 21, 23, 28–30, 32, 34, 52, 66, 67, 92
thermoplastic 9, 28, 65, 99, 139, 148, 159
–, amorphous 118
– filament 146, 147
– matrices 108, 112, 118, 126, 137, 139, 140, 144–146
– parts 148
– processing technology 147, 148
– resins 9, 17, 18, 23, 24, 39, 41, 42–57, 100, 101, 109, 110, 160, 161, 164, 168, 173, 175, 176, 178, 179
Thermoplastic Composites 110, 132, 140, 144–155, 171
–, CF reinforced 144
thermoset 38, 65, 99, 110, 139, 146–148, 159, 164, 176, 178, 179
–, tough 137, 138
thermoset matrices 9, 108, 112, 139, 140, 146, 174, 178
–, damage tolerant 118, 138
thermosetting resins 10–42
thermotropic 57, 58
thick
– structures 142
– wall 176
thin structure 118, 136
thiol 14
Thornel 89–92, 129, 139, 149, 151–154, 168, 171, 175
TICA 34
Titan 167, 168
titanate coupling agents 110
titanates, organo 109
titanium 170, 174
three dimensional, 3D 140
TLC 99
Tm 44, 46, 81
TMA 47, 100
toluene 45
toluene diisocyanate 53
o-toluidine 36
tool life 145, 146
tooling 146, 148, 177–179
–, rubber punch 148
topological defects 79
Torlon 44, 54, 159
torque 172, 180, 182
tough 29, 39, 49
toughened 140
– resin composites 137
tougher 28, 29, 38, 56, 60, 66,
toughness 17, 18, 26, 29, 38, 39, 42, 43, 45, 49, 52–54, 57, 66, 75, 86, 94, 102,

110, 111, 120, 125, 126, 138–140, 144, 145, 159, 171, 174, 181, 183
tow 48, 90, 147, 168, 179
–, commingled 147
–, unidirectional 147
toxicity 18, 102, 176
toxicology 176
trains 183
transmissions 20, 180, 182
transverse
– composite modulus 120, 121
– compressive strength 122
– direction 89, 91, 115, 116, 165
– inner plies 124
– properties 91, 117, 123, 127
– strength 123
– tensile strength 122, 127
– tensile strain, ultimate 137
tri-arch bridge 185
triaryl phosphine 21
triazine 41
triblock 64
trimellitoyl acid chloride 54
trimerization 39
trimethyl ester 26
trucks 7
TTT 10, 11, 100
tubes 181
Tungsten 21
turbine blades 105
turbochargers 180
Twaron 82, 84, 85
twisting 173, 174
twists 79

Udel 44
UDRI 129
UHMWPE 75, 76, 78–81, 93, 94, 109, 110
ultimate strain 126, 136
ultra drawing 80
ultrasonic C scan 103
ultrasonic waves 77
ultrasound 103
unbounded 106
uncyclized 37
uniaxial 89
– composites 132, 134
unidirectional 48, 78, 89, 116, 146, 173
unifabric 148
unitape 148–150, 155
unsaturated 23
USS Essex 176, 177
unstable 24
untangling 79
UV 58, 80

V-22 4, 5
V_{50} 164, 165
vacuum bag 161
vapor deposition 110
Vectra 57, 58, 145
vehicle 7, 23, 93, 94, 102, 164, 167, 182
–, lift 167
vehicles, land 164, 166
velocity 172, 174
ventures 8
vessels 102, 166, 167
–, U.S. Navy 93, 94
–, pressure 93, 104, 168, 170
vests 94
vibration 4, 20, 106, 167, 172, 173, 174
– analysis 106
– dampening 181
–, mechanical 182
vinyl 32
– esters 11, 164, 185
vinyls 139
violins 184
viscosity 10, 20, 24, 82, 146
–, inherent 45, 48, 50
–, intrinsic 82
–, melt 10, 39, 48
vitrification 10
void
– content 64, 78, 103, 123
– free 147
– level 121
voids 123, 129, 142, 146, 179
volatile 20, 22, 25, 39
volume 85
volume fraction 1
–, void 123

walls, galley and lavatory 160
warpage 22, 163
water 14, 15, 20, 24, 26, 85, 99, 112
– immersion 103
– sensitive 40, 82
– sensitivity 17
– skis 175
– spray 103
water sorption 82, 85, 99, 139, 155

–, low equilibrium 138
WAXS 64
weak locations 108
weaknesses 8
wear 162, 181
– resistance 45
– resistant 163
weather resistance 114
weave style 146, 147
weaving 62, 109
Weibull distribution 112
weight 5, 34, 52, 85, 88, 167, 168, 170, 171, 173–176, 178–180, 183
– loss 28, 34, 52
–, lower 1
– savings 157, 163, 167, 170, 176, 179
wet
– conditions 123, 127, 131, 132, 134, 155
– performance 127, 140
– service 141, 142, 144, 149
– state 123
welding 168
–, seam fusion 148
wheels 173, 174
window
– frames 160
– openings 184
wind tunnels 173
wing(s) 53, 157, 159
wing skins, airplane 118
wood 4, 172, 173, 175, 176, 185
wooden 174, 175
woven 137, 146, 147, 167, 179, 185

XPS 85, 89, 97, 98, 108–110
XPS, grazing angle 85
X-Ray 105, 181
– Diffraction 91, 101
– Radiography 105
– technology 105
Xydar 57, 58, 145
xylene 79
o-xylylene 33
p-xylylene 33

yarns 146, 147
yield strength 129
Young's modulus 77

zircoaluminates 109
zirconates 109
zirconium 21

J.H. Fendler

Membrane-Mimetic Approach to Advanced Materials

1994. X, 236 pp. 134 figs., 11 tabs. (Advances in Polymer Science, Vol. 113)
ISBN 3-540-57237-6

Contents: Preparation and Characterization of Compartments.- Metallic and Catalytic Particles.- Semiconductor Particles and Particulate Films.- Conductors and Superconductors.- Magnetism, Magnetic Particles, and Magnetic Particulate Films in Membrane-Mimetic Compartments.- Advances Ceramics.

P. Dubin, J. Bock, R. Davis, D.N. Schulz, C. Thies (Eds.)

Macromolecular Complexes in Chemistry and Biology

1994. XX, 359 pp. 196 figs., 39 tabs. ISBN 3-540-57166-3

The current book describes the chemical and physical behavior of polymers and biopolymers that form highly associating structures in equilibrium solution. It summons the established results known of polymer complexes in solution taking into account also the recent developments in biotechnology concerning this topic, in technological applications of polymer-protein interactions, in fluorescence and scattering techniques for the study of intra- and interpolymer association and in the study of ionomers in solution. The book covers the whole range from synthesis and fundamental aspects to applications and technology of associated polymers.